107 Geometry Problems

From the AwesomeMath Year-Round Program

107 Geometry Problems

From the AwesomeMath Year-Round Program

Titu Andreescu

Michal Rolínek

Josef Tkadlec

Library of Congress Control Number: 2012920775

ISBN-13:978-0-9799269-7-6 ISBN-10:0-9799269-7-1

9 8 7 6 5 4 3 2 1

www.awesomemath.org

Cover design by Iury Ulzutuev

Preface

This book is a sequel to *106 Geometry Problems from the AwesomeMath Summer Program.* It contains 107 geometry questions used in the AwesomeMath Year-Round Program which trains and tests top middle and high-school students from U. S. and around the world.

The book begins with a theoretical chapter, where we review basic facts and familiarize the reader with some more advanced techniques. We then proceed to the main part of the work, the problem sections. The problems are a carefully selected and balanced mix which offers a vast variety of flavors and difficulties, ranging from AMC and AIME levels to high-end IMO problems. Out of thousands of Olympiad problems from around the globe we chose those which best illustrate the featured techniques and their applications. The problems meet our demanding taste and fully exhibit the enchanting beauty of classical geometry. For every problem we provide a detailed solution and strive to pass on the intuition and motivation lying behind. Numerous problems have multiple solutions.

Directly experiencing Olympiad geometry both as contestants and instructors, we are convinced that a neat diagram is essential to efficiently solving a geometry problem. Our diagrams do not contain anything superfluous, yet emphasize the key elements and benefit from a good choice of orientation. Many of the proofs should be legible only from looking at diagrams.

In the theoretical part we discuss some advanced theorems from triangle geometry and develop the theory of transformations, such as homothety, spiral similarity, and inversion. Employing the latter, we demonstrate the effectiveness of dynamic geometric thinking.

True mastery of geometry relies on proficient use of common sense. Therefore, we chose to avoid analytical and computational techniques such as complex numbers, vectors, or barycentric coordinates.

Although the primary audience for this book consists of high-performing students and their teachers, anyone with an interest in Euclidean geometry or recreational mathematics is invited to join this geometric excursion.

Finally, we would like to express our gratitude to Richard Stong and Cosmin Pohoaţă for critiquing the entire manuscript and providing fruitful comments.

We wish you a pleasant reading.

The Authors

Abbreviations and Notation

Notation of geometrical elements

$\angle BAC$	convex angle by vertex A
$\angle(p,q)$	directed angle between lines p and q
$\angle BAC \equiv \angle B'AC'$	angles BAC and $B'AC'$ coincide
AB	line through points A and B, distance between points A and B
$\overline{AB}$	directed segment from point A to point B
$X \in AB$	X lies on the line AB
$X = AC \cap BD$	X is the intersection of the lines AC and BD
$\triangle ABC$	triangle ABC
$[ABC]$	area of $\triangle ABC$
$[A_1 \dots A_n]$	area of polygon $A_1 \dots A_n$
$AB \parallel CD$	lines AB and CD are parallel
$AB \perp CD$	lines AB and CD are perpendicular
$p(X,\omega)$	power of point X with respect to circle ω
$\triangle ABC \cong \triangle DEF$	triangles ABC and DEF are congruent (in this order of vertices)
$\triangle ABC \sim \triangle DEF$	triangles ABC and DEF are similar (in this order of vertices)
$\mathcal{H}(H,k)$	homothety with center H and factor k
$\mathcal{S}(S,k,\varphi)$	spiral similarity with center S, dilation factor k, and angle of rotation φ

Notation of triangle elements

a, b, c	sides or side lengths of $\triangle ABC$
$\angle A, \angle B, \angle C$	angles by vertices A, B, and C of $\triangle ABC$
s	semiperimeter
x, y, z	expressions $\frac{1}{2}(b+c-a)$, $\frac{1}{2}(c+a-b)$, $\frac{1}{2}(a+b-c)$
r	inradius
R	circumradius
K	area
h_a, h_b, h_c	altitudes in $\triangle ABC$
m_a, m_b, m_c	medians in $\triangle ABC$
l_a, l_b, l_c	angle bisectors (segments) in $\triangle ABC$
r_a, r_b, r_c	exradii in $\triangle ABC$

Abbreviations

AMC10	American Mathematics Contest 10
AMC12	American Mathematics Contest 12
AIME	American Invitational Mathematics Examination
USAMTS	United States of America Mathematical Talent Search
USAJMO	United States of America Junior Mathematical Olympiad
USAMO	United States of America Mathematical Olympiad
USA TST	United States of America IMO Team Selection Test
MEMO	Middle European Mathematical Olympiad
IMO	International Mathematical Olympiad
Putnam	William Lowell Putnam Mathematical Competition

Contents

Chapter 1

Advanced Topics in Geometry

Overview of Basic Techniques

Let us begin with reviewing some basic facts and techniques. Knowing them is not essential for further reading so don't get discouraged if you have gaps now and then. On the other hand, in order to learn the most from this book, we strongly recommend to get a firm grasp of what is presented in this section. All proofs (and much more) can be found in the preceding book *106 Geometry Problems from the AwesomeMath Summer Program.*

First Triangle Centers

Proposition 1.1 (Existence of the circumcenter). *In triangle ABC the perpendicular bisectors of AB, BC, and CA meet at a single point. This point is called the circumcenter of triangle ABC, is usually denoted by O, and it is the center of the circumscribed circle (or simply circumcircle).*

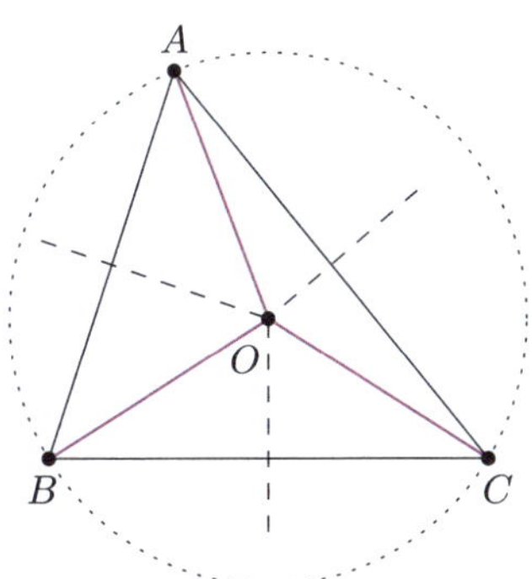

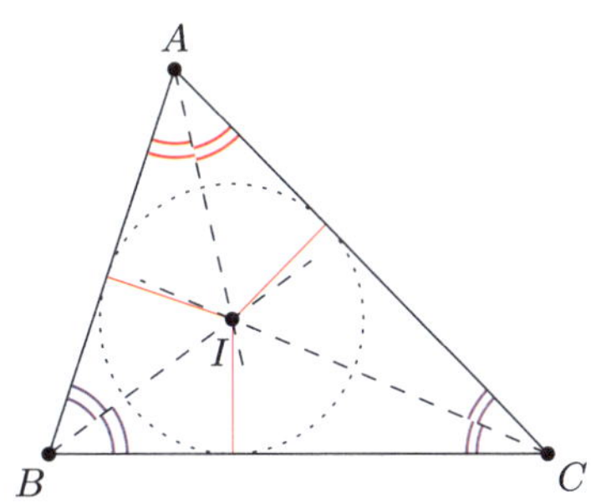

Proposition 1.2 (Existence of the incenter). *In triangle ABC the internal angle bisectors meet at a point. This point is called the incenter of triangle*

ABC, is usually denoted by I, and it is the center of the incircle of triangle ABC.

Proposition 1.3 (Existence of the orthocenter). *In triangle ABC the altitudes meet at a single point. This point is called the orthocenter of triangle ABC and is usually denoted by H.*

Proposition 1.4 (Existence of the centroid). *In triangle ABC the medians meet at a point. This point is called the centroid of triangle ABC and is usually denoted by G.*

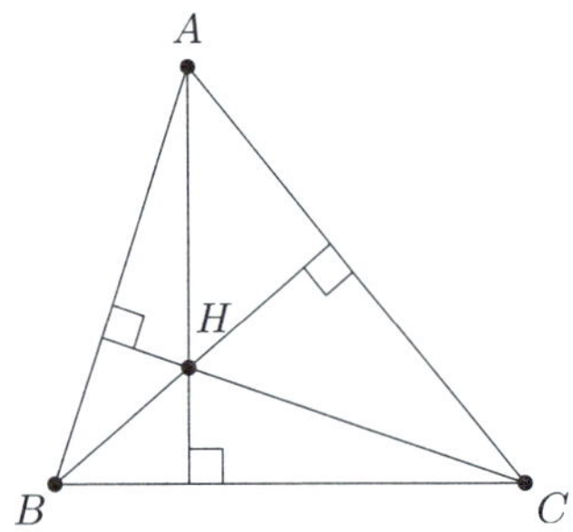

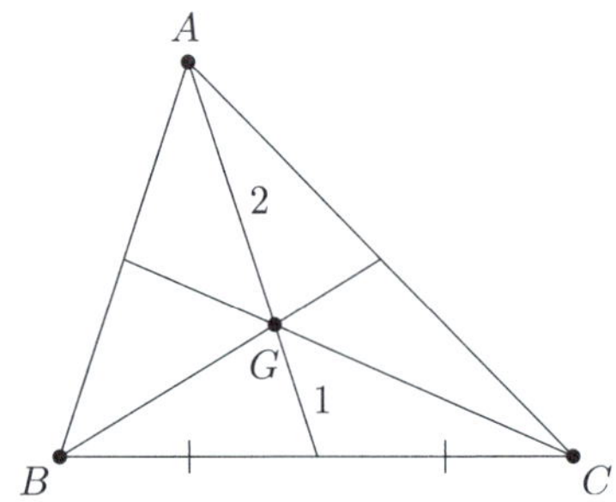

Proposition 1.5 (Existence of the excenter). *In triangle ABC the A-angle bisector and the bisectors of external angle B and C meet at a point. This point is called the A-excenter of triangle ABC, is usually denoted by I_a and it is the center of the A-excircle (circle tangent to the side BC and to the extended sidelines AB and AC). Similarly, we define points I_b and I_c.*

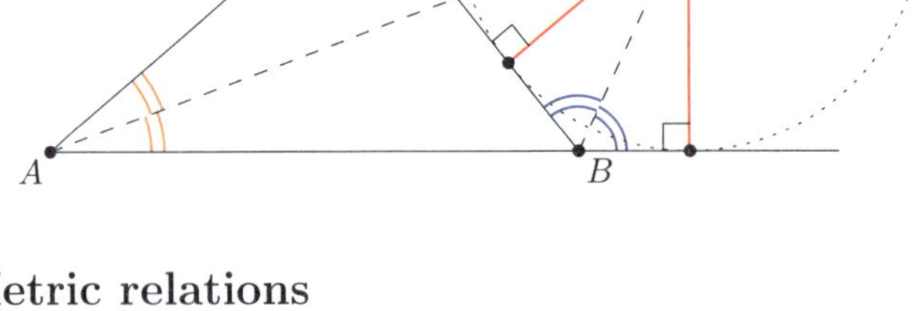

Metric relations

Proposition 1.6 (Equal Tangents). *Two tangent lines to the given circle ω intersect at A. Denote by B, C the points of tangency with the circle. Then AB = AC.*

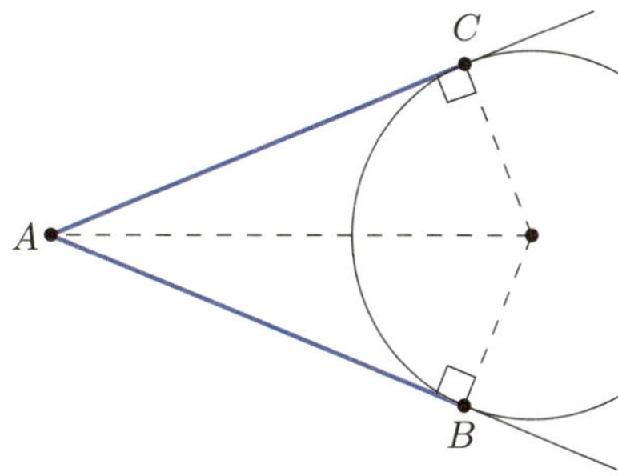

We use the following standard xyz notation in triangle ABC with semiperimeter s:

$$x = s - a = \frac{1}{2}(b+c-a), \quad y = s - b = \frac{1}{2}(c+a-b), \quad z = s - c = \frac{1}{2}(a+b-c),$$

the purpose of which is revealed in the next two propositions.

Proposition 1.7 (Points of contact). *Let ABC be a triangle with semiperimeter s. Denote by D, E, F the points of tangency of the incircle with the sides BC, CA, AB, respectively. Also let the A-excircle touch the lines BC, CA, AB at points K, L, M, respectively. Then the following hold:*

(a) $AE = AF = x, \quad BD = BF = y, \quad CD = CE = z.$
(b) $AL = AM = s.$
(c) Points K and D are symmetric with respect to the midpoint of BC.

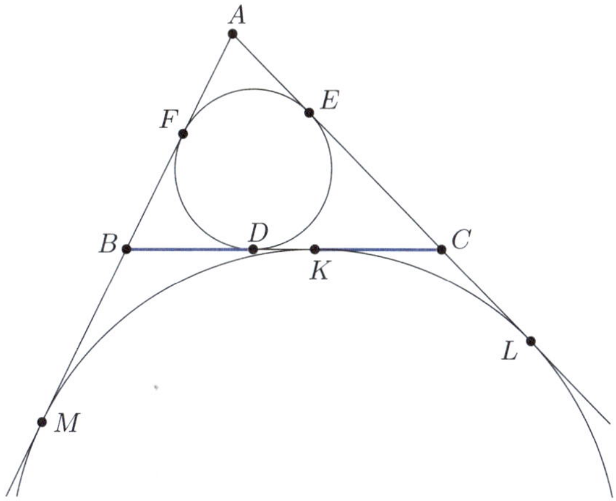

Proposition 1.8 (xyz formulas). *In triangle ABC we can find the area K, inradius r, and circumradius R in terms of x, y, z as follows:*

(a)

$$K = \sqrt{(x+y+z)xyz},$$

(b)
$$r=\sqrt{\frac{xyz}{x+y+z}},$$

(c)
$$R=\frac{(y+z)(z+x)(x+y)}{4\sqrt{xyz(x+y+z)}}.$$

Theorem 1.9 (The Extended Law of Sines). *Let ABC be a triangle. Then*

$$\frac{a}{\sin\angle A}=\frac{b}{\sin\angle B}=\frac{c}{\sin\angle C}=2R,$$

where R is the circumradius of triangle ABC.

Theorem 1.10 (Angle Bisector Theorem). *In triangle ABC let AD, $D\in BC$, be the internal angle bisector. Then*

$$\frac{BD}{CD}=\frac{c}{b},\quad BD=\frac{ac}{b+c},\quad CD=\frac{ab}{b+c}.$$

Theorem 1.11 (The Law of Cosines). *Let ABC be a triangle. Then*

$$a^2=b^2+c^2-2bc\cos\angle A.$$

Circles, Tangents

Theorem 1.12 (Inscribed Angle Theorem). *Let BC be a chord of a circle ω centered at O and let $A\in\omega$, $A\neq B,C$. Then the inscribed angle BAC corresponding to arc BC equals one half of the central angle corresponding to the same arc.*

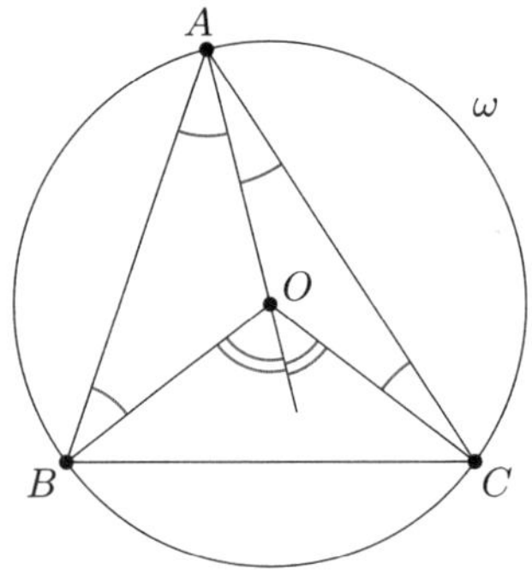

Quadrilaterals which are inscribed in a circle are called *cyclic* and play fundamental role in the technique called *angle-chasing*.

Proposition 1.13 (The key properties of cyclic quadrilaterals). *Let $ABCD$ be a convex quadrilateral. Then:*

(a) If $ABCD$ is cyclic then any of its sides is visible from the other two vertices under the same angle, and any of its diagonals is visible from the other two vertices under angles that sum up to $180°$.
(b) If there is a side of $ABCD$ that is visible from the other two vertices under the same angle, then $ABCD$ is cyclic.
(c) If there is a diagonal of $ABCD$ that is visible from the other two vertices under angles that sum up to $180°$, then $ABCD$ is cyclic.

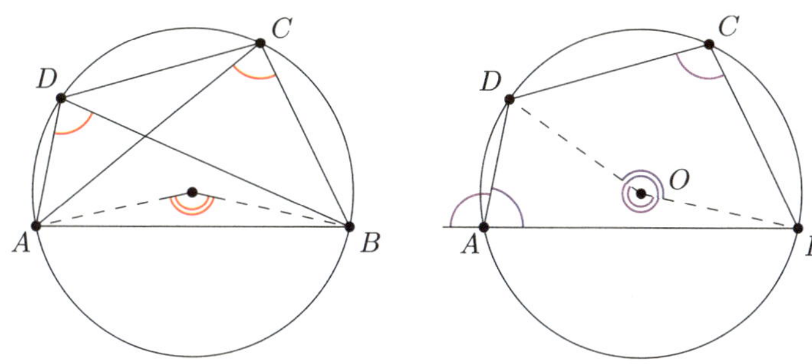

Corollary 1.14 (Angle between chords or secants). *Let $ABCD$ be a quadrilateral inscribed in a circle ω and denote by P the intersection of its diagonals. Suppose that rays BA and CD intersect at R. Finally, denote the inscribed angles corresponding to arcs BC, DA (not containing A, B) by β, δ. Then*

(a) $\angle BPC = \beta + \delta$,
(b) $\angle BRC = \beta - \delta$.

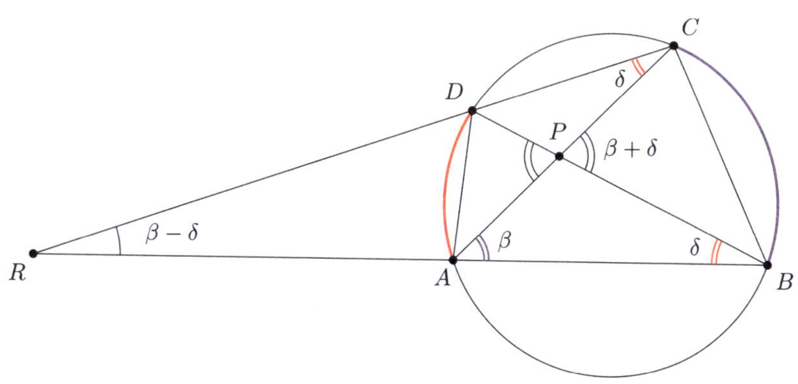

Proposition 1.15 (Angle by tangent). *Let ABC be a triangle inscribed in a circle ω. Let ℓ be a line passing through A different from AB. Let L be a*

point on ℓ such that AB separates points C, L. Then AL is tangent to ω if and only if $\angle LAB = \angle ACB$.

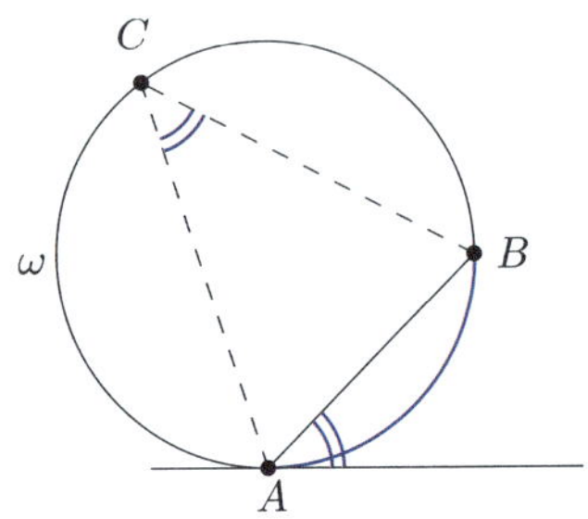

Antiparallel lines

Given a line n we say that lines ℓ and m (neither parallel to n) are **antiparallel** with respect to line n if the reflection ℓ' of ℓ about n is parallel to m. Observe that the following holds:

(a) If ℓ is antiparallel to m then it is antiparallel to all lines parallel to m.
(b) (Symmetry) If ℓ is antiparallel to m then m is antiparallel to ℓ.
(c) Given a line n and a set of mutually parallel lines, then lines antiparallel to all of these with respect to n form again a set of mutually parallel lines.

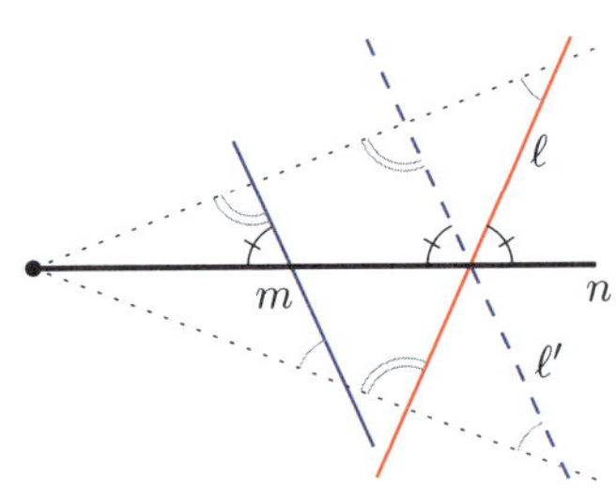

Proposition 1.16. *Let line m intersect rays OA, OB of angle AOB at distinct points X, Y, respectively. Let line ℓ, ($\ell \neq m$) intersect lines OA, OB of angle AOB at (not necessarily distinct) points P, Q, respectively. Then ℓ and m are antiparallel with respect to the angle bisector of angle AOB if and only if one of the following (based on the configuration) holds:*

(a) Points X, Y, P, Q are concyclic (if they are pairwise distinct).
(b) Line OA is tangent to the circumcircle of triangle XYQ (if $X = P$). A similar result holds if $Y = Q$.

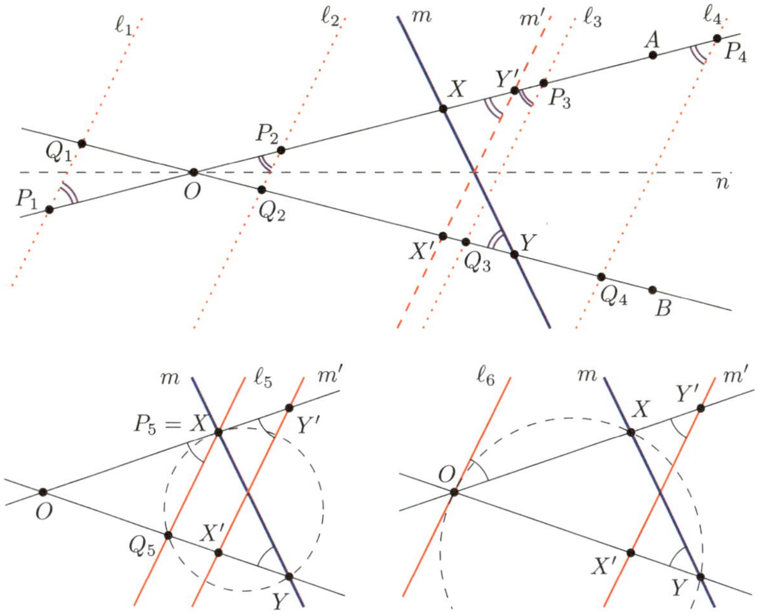

(c) *Line ℓ is tangent to the circumcircle of triangle XYO (if ℓ passes through O).*

Since antiparallel lines are usually taken with respect to the angle bisector of some angle, let us in that case call these lines *antiparallel with respect to* that angle or simply *antiparallel in* that angle. Of particular interest are antiparallel lines that both pass through the vertex of an angle – such lines are called *isogonal*. One pair of isogonal lines is especially worth emphasizing.

Proposition 1.17 (H and O are friends)**.** *In triangle ABC points H (the orthocenter) and O (the circumcenter) lie on isogonal lines in each of the angles $\angle A$, $\angle B$, $\angle C$.*

Directed angles mod[1] $180°$

The magnitude of an angle between lines l, m intersecting at vertex O can be viewed as a number from interval $[0, 180)$ describing (in degrees) the amount of counter-clockwise rotation around O which takes l to m. Let us call this quantity **the directed measure of an angle** and denote it by $\angle(l, m)$. Note that order of lines in brackets matters – in fact $\angle(l, m) + \angle(m, l) = 180°$. This notion will be our main weapon for simplifying angle-chasing casework throughout the book.

[1] This means, we shall work with remainders after division by 180. For example, instead of $200°$, we shall work with $20°$.

Proposition 1.18. *(a)* $\angle(l,m)+\angle(m,n)=\angle(l,n)$, *with addition mod* $180°$.

(b) For any point P $\angle(PA,AB)=\angle(PA,AC)$ *if and only if points* A, B, C *lie on a single line in some order.*

(c) $\angle(AC,CB)=\angle(AD,DB)$ *if and only if points* A, B, C, D *lie on one circle in some order.*

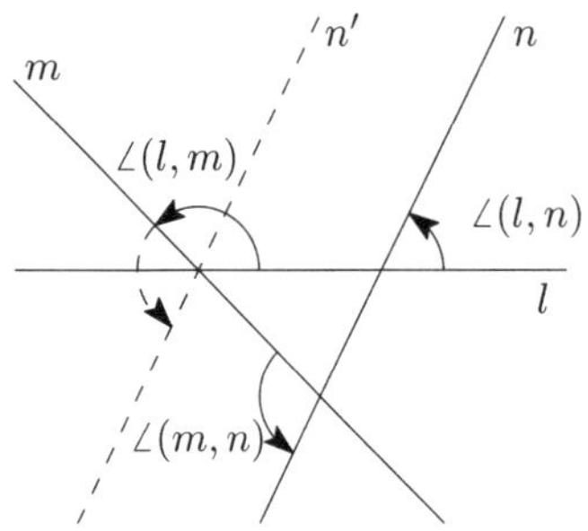

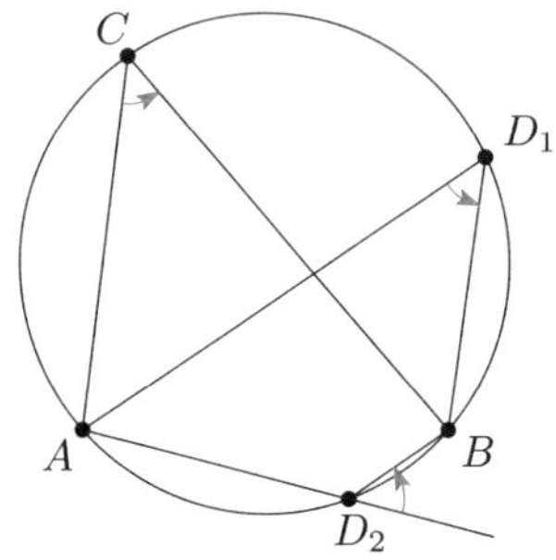

Power of a Point

Proposition 1.19. *(a) Let* $ABCD$ *be a convex quadrilateral and let* $P=AC\cap BD$. *Then the points* A, B, C, D *are concyclic if and only if*

$$PC\cdot PA=PB\cdot PD.$$

(b) Let $ABCD$ *be a convex quadrilateral and let* $P=AB\cap CD$. *Then the points* A, B, C, D *are concyclic if and only if*

$$PA\cdot PB=PC\cdot PD.$$

(c) Assume points P, B, C *are collinear in this order and point* A *does not lie on this line. Then the line* PA *is tangent to the circumcircle of triangle* ABC *if and only if*

$$PA^2=PB\cdot PC.$$

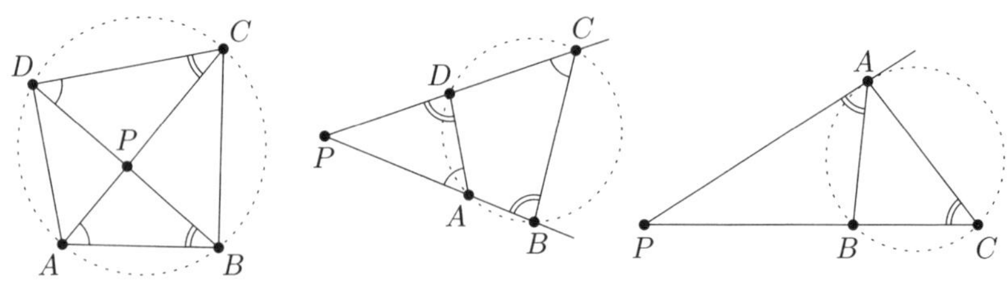

Theorem 1.20 (Power of a Point). *Given point* P *and circle* ω, *let* ℓ *be an arbitrary line passing through* P *and intersecting* ω *at points* A *and* B. *Then*

the value of $PA \cdot PB$ does not depend on the choice of ℓ. Also, if P lies outside of ω and PT, $T \in \omega$, is a tangent to ω then $PA \cdot PB = PT^2$.

If we denote the center of ω by O and its radius by R then $PA \cdot PB = |OP^2 - R^2|$. The quantity

$$p(P, \omega) = OP^2 - R^2$$

is called the power of point P with respect to circle ω.

Note that the number $p(P, \omega)$ is negative when P lies inside ω, zero when it lies on ω, and positive otherwise.

Proposition 1.21 (Radical axis). *Let ω_1, ω_2 be two circles with distinct centers O_1, O_2 and radii R_1, R_2, respectively. Then the locus of points X for which $p(X, \omega_1) = p(X, \omega_2)$ is a line perpendicular to O_1O_2. This line is called the radical axis of the two circles.*

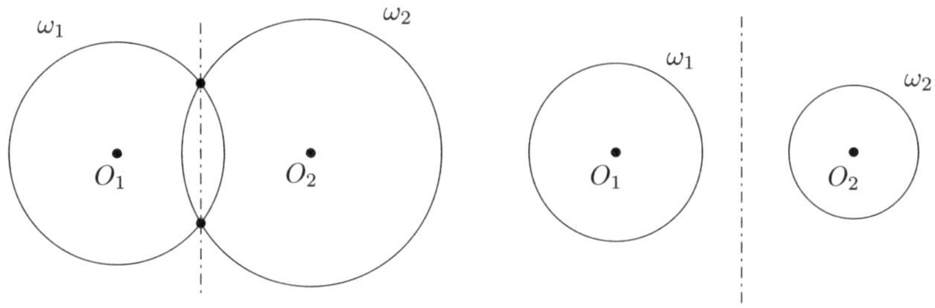

The radical axis is a powerful tool in many problems involving intersecting circles since in that case the radical axis is the line joining their intersections, which both have equal (namely zero) power with respect to the two circles.

Proposition 1.22 (Radical center). *Let $\omega_1, \omega_2, \omega_3$ be circles with pairwise distinct centers. Then their pairwise radical axes are either parallel or concurrent. The point of concurrence is called the radical center of the three circles.*

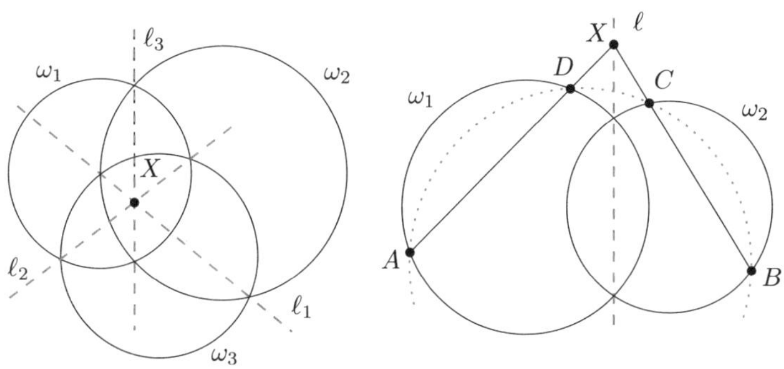

Proposition 1.23 (Radical Lemma). *Let line ℓ be radical axis of the circles ω_1, ω_2. Let A, D be distinct points on ω_1 and let B, C be distinct points on ω_2 such that the lines AD and BC are not parallel. Then the lines AD and BC intersect at ℓ if and only if $ABCD$ is cyclic.*

Theorem 1.24 (Menelaus'[2] Theorem). *Let ABC be a triangle and let points D, E, F lie on the lines BC, CA, AB, respectively, so that either none or two of them lie on the triangle sides. Then the points D, E, F are collinear if and only if*

$$\frac{BD}{DC} \cdot \frac{CE}{EA} \cdot \frac{AF}{FB} = 1.$$

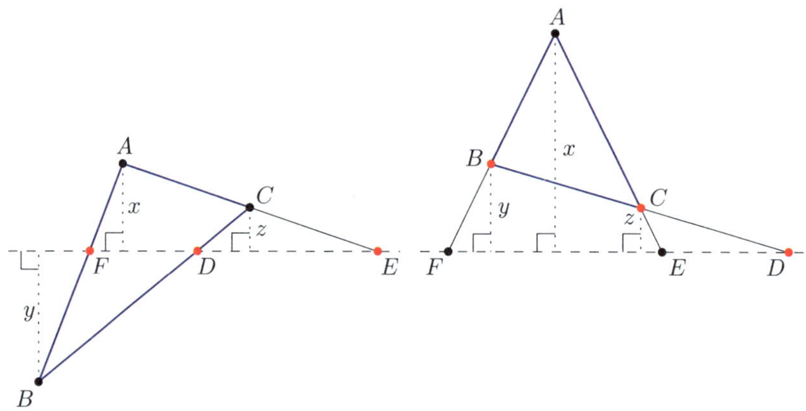

Segments which connect vertex of a triangle with a point on the opposite side are called *cevians*.

Theorem 1.25 (Ceva[3]'s Theorem). *Let ABC be a triangle, and let P, Q, R be points on the sides BC, CA, AB, respectively. Then the lines AP, BQ, CR are concurrent if and only if*

$$\frac{BP}{PC} \cdot \frac{CQ}{QA} \cdot \frac{AR}{RB} = 1.$$

Theorem 1.26 (Existence of isogonal conjugate). *Let cevians AP, BQ, CR concur at point X. Now construct cevians AP', BQ', CR' which are isogonal to AP, BQ, CR, respectively, in the respective angles. Then the cevians AP', BQ', CR' are concurrent. The point of concurrence is called the isogonal conjugate of X.*

[2] Menelaus of Alexandria (c. 70–140) was a Greek mathematician and astronomer.
[3] Giovanni Ceva (1647–1734) was an Italian mathematician.

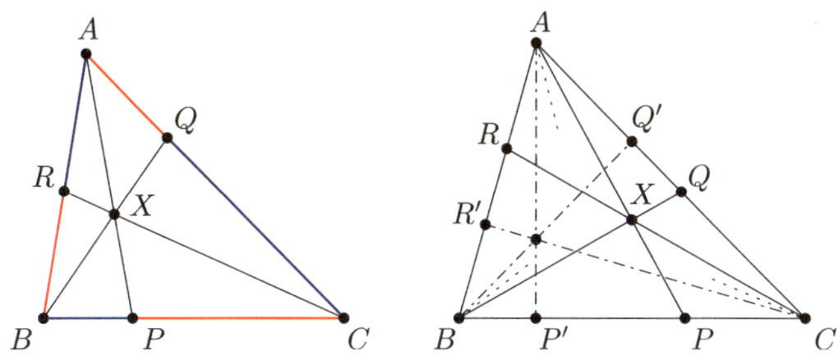

Directed segments

A **directed segment** emanating from A with endpoint B will be denoted by $\overline{AB}$.

The important property of directed segments is that the ratio or the product of two directed segments, which are part of the same line, is assigned a sign. The sign is positive if the directed segments have the same orientation and negative otherwise. By the same logic we have

$$\overline{AB} = -\overline{BA}.$$

Homothety

It is our everyday experience that if we zoom onto certain point, objects don't change shape, only size. In this section we give mathematical background to the idea of scaling and reveal its further properties.

Given a point H and a number k different from 0 and 1, *homothety* (or *dilation*) with center H and factor k is the transformation of the plane which sends point A to a point A' such that:

(a) Points H, A and A' are collinear.
(b) $\overline{HA'} = k \cdot \overline{HA}$.

We denote such homothety by $\mathcal{H}(H,k)$.

Part (b) can be equivalently stated without using directed segments if one adds that for $k > 0$ the rays HA and HA' coincide and for $k < 0$ they are mutually opposite.

Observe that choice $k = -1$ corresponds to point reflection.

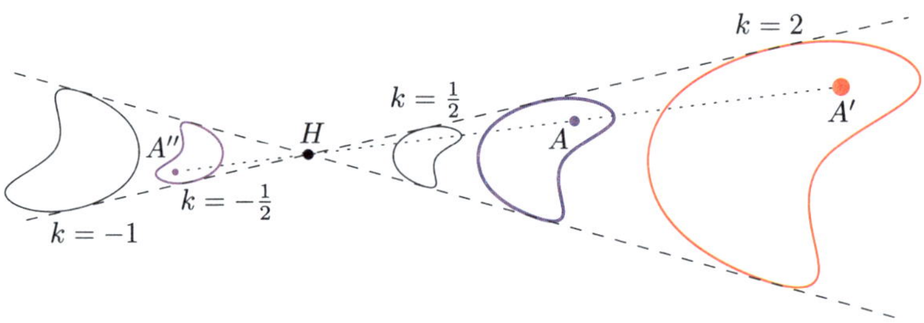

Proposition 1.27. *Let $\mathcal{H}(H,k)$ be a homothety and denote the images of distinct non-collinear points A, B, C by A', B', C', respectively. Then:*

(a) Line $A'B'$ is parallel to AB. Moreover, $A'B' = k \cdot AB$.
(b) Homothety preserves angles and ratios. In other words, $\angle A'B'C' = \angle ABC$ and

$$\frac{A'B'}{B'C'} = \frac{AB}{BC}.$$

Proof. (a) If the points H, A, B are collinear, the proposition is valid trivially. Otherwise, note that as $HA' = k \cdot HA$, $HB' = k \cdot HB$, and $\angle A'HB' = \angle AHB$, by SAS we have $\triangle AHB \sim \triangle A'HB'$ with factor k, so $AB \parallel A'B'$ and $A'B' = k \cdot AB$.

(b) Since $A'B' \parallel AB$ and $B'C' \parallel BC$, we have $\angle A'B'C' = \angle ABC$. Also

$$\frac{A'B'}{B'C'} = \frac{k \cdot AB}{k \cdot BC} = \frac{AB}{BC},$$

which proves the second part.

□

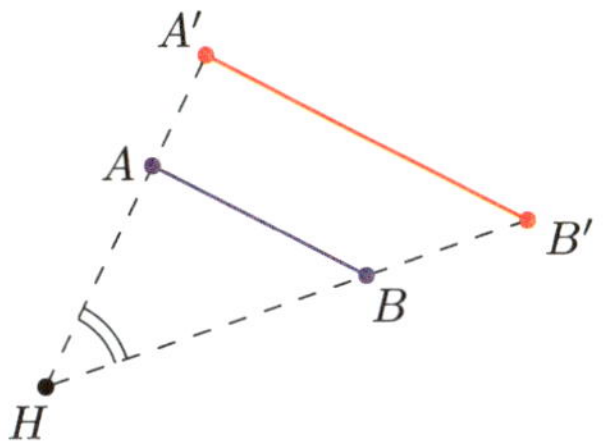

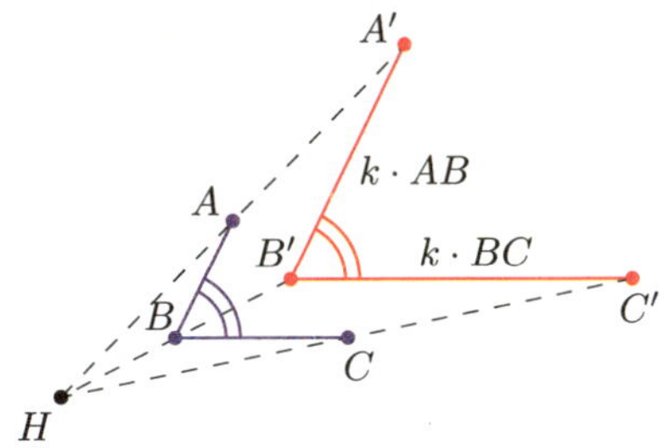

Since we have proved that homothety preserves angles, ratios and directions, we may now state (leaving details for the reader) that the image of a figure is a similar figure of the same orientation. In particular:

(a) The image of a line is a parallel line.
(b) The image of a triangle is a similar triangle with corresponding sides parallel.
(c) The image of a circle is a circle.

Proposition 1.28. *(a) Given two parallel segments AB and $A'B'$ of different length, there exists unique homothety that maps A to A' and B to B'.*
(b) Let ABC and $A'B'C'$ be two non-congruent triangles with parallel corresponding sides. Then there exists unique homothety that maps triangle ABC to triangle $A'B'C'$. As a result, lines AA', BB', CC' are concurrent.

Proof. (a) First note that the center of such homothety has to lie on the lines AA' and BB' and denote their intersection by H. Now triangles HAB and $HA'B'$ are similar (AA), so $HA'/HA = HB'/HB$ and homothety $\mathcal{H}(H, HA'/HA)$ does the job (the case when all the points are collinear is left to the reader as a boring algebra exercise).
(b) Denote by H the center of homothety that maps AB to $A'B'$.

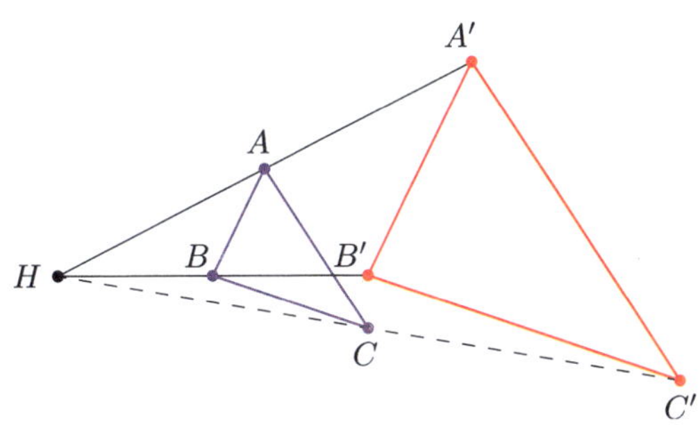

Such homothety sends triangle ABC to some triangle $A'B'X$. Since both triangles $A'B'X$ and $A'B'C'$ are similar to triangle ABC and have the same orientation, they are in fact identical, and hence H, C and C' are collinear. □

Keeping these properties of homothety in mind we are now ready to solve some examples.

Example 1.1 (Tournament of Towns 1984). *Let P be a point inside a given square $ABCD$. Prove that the centroids of triangles ABP, BCP, CDP, DAP form a square.*

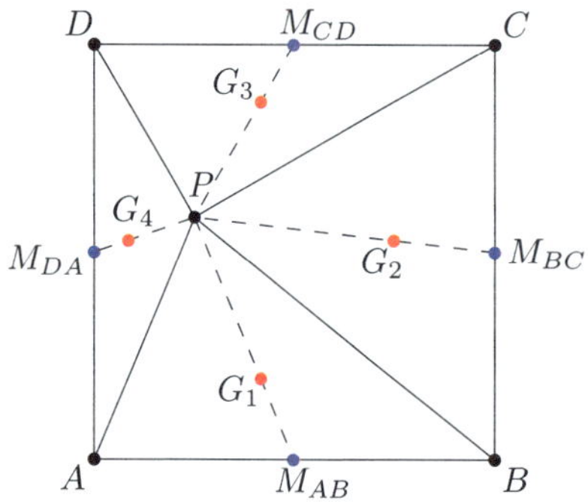

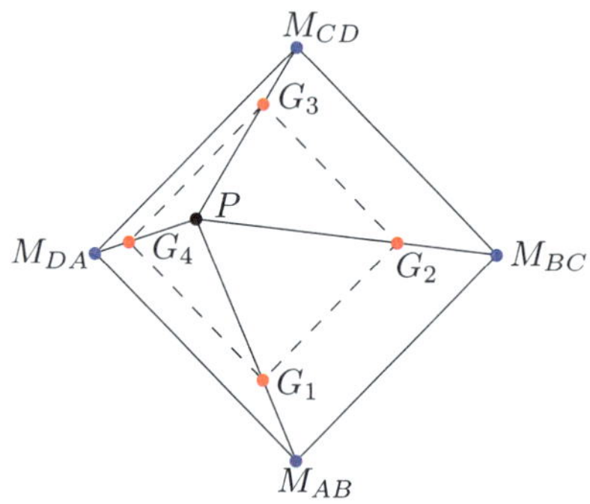

Proof. Denote the centroids by G_1, G_2, G_3, G_4, respectively, and the midpoints of the respective sides of $ABCD$ by M_{AB}, M_{BC}, M_{CD}, M_{DA}. Since the centroid divides the median of a triangle in ratio $2:1$, a homothety $\mathcal{H}(P, \frac{2}{3})$ sends quadrilateral $M_{AB}M_{BC}M_{CD}M_{DA}$ to the quadrilateral $G_1G_2G_3G_4$. As $M_{AB}M_{BC}M_{CD}M_{DA}$ is a square, $G_1G_2G_3G_4$ is also a square. $\square$

Example 1.2. *Let $ABCD$ be a trapezoid with $AB \parallel CD$ and denote by E the intersection of its diagonals. Construct equilateral triangles ABF, CDG externally. Prove that points E, F, G are collinear.*

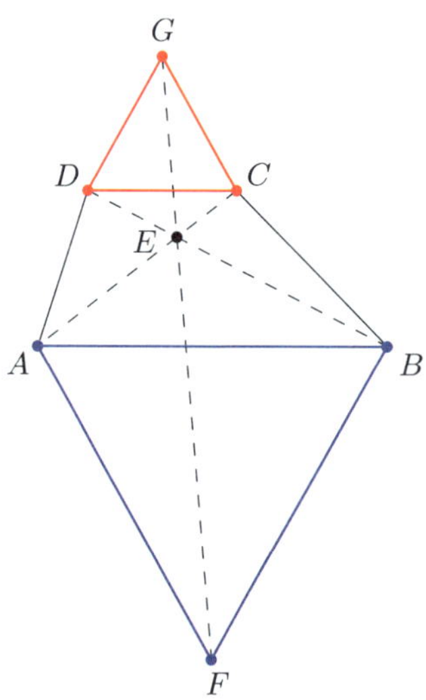

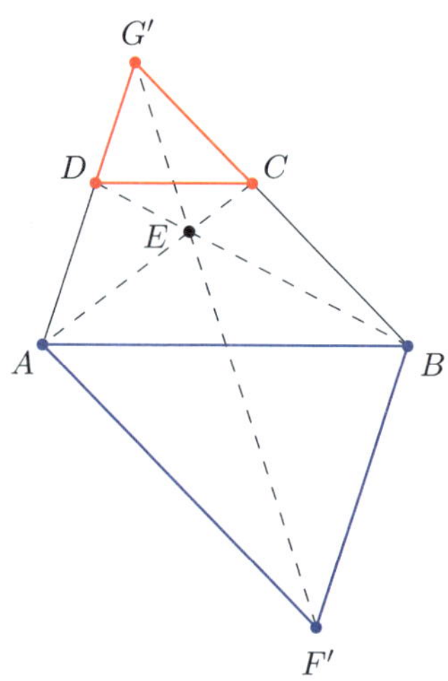

Proof. As triangles ABF and CDG are similar and have parallel corresponding sides, there exists a homothety $\mathcal{H}$ that maps triangle ABF to triangle CDG. Thus by Proposition 1.28(b) the lines AC, BD and FG are concurrent at the center of this homothety implying that E lies on the line FG. □

Remark. It is good to realize that the previous argument is not specific to the shape of an equilateral triangle. Any pair of (directly) similar triangles DCG and BAF would do the trick. One important example is the choice of G' as intersection of AD and BC (see the diagram on the right). Then F' completes the parallelogram $G'AF'B$ and the conclusion reads as E lies on the G'-median of the triangle ABG'.

The following example reveals an important fact from triangle geometry.

Example 1.3 (Euler[4] line). *Let ABC be a non-equilateral triangle and let H, G, O be its orthocenter, centroid, and circumcenter, respectively. Then the points H, G, O lie on a single line (called the Euler line of triangle ABC) in this order, and $HG = 2 \cdot GO$.*

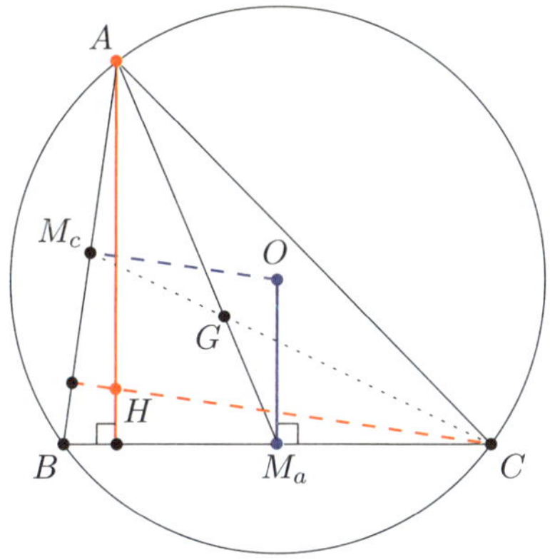

Proof. Denote by M_a, M_c the midpoints of sides BC, AB, respectively, and consider homothety $\mathcal{H}(G, -2)$.

Since the centroid divides the median in ratio $2:1$, the image of M_a under $\mathcal{H}$ is A. Also as every homothety maps a line to a parallel line, $\mathcal{H}$ sends the perpendicular bisector OM_a to the A-altitude of triangle ABC.

By exactly the same argument we find out that $\mathcal{H}$ sends line OM_c to the C-altitude. Therefore it sends the intersection of lines OM_a and OM_c (which is O) to the intersection of A-altitude and C-altitude (which is H). Hence points O, G, H are collinear and satisfy

$$\overline{GH} = -2 \cdot \overline{GO},$$

as desired. □

[4]Leonhard Euler (1707–1783) was a Swiss mathematician and physicist.

Homothety is also a powerful instrument when dealing with circles. Especially, if they are mutually tangent.

Proposition 1.29. *Let ω_1, ω_2 be circles of different radii r_1, r_2 centered at O_1, O_2, respectively.*

(a) There exist two homotheties, one (call it $\mathcal{H}^+$) with positive factor and the other (call it $\mathcal{H}^-$) with negative factor, that map ω_1 to ω_2.

(b) If common external tangents of ω_1 and ω_2 exist and intersect at H^+, then H^+ is the center of homothety $\mathcal{H}^+$. Similarly, if common internal tangents of ω_1, ω_2 exist and intersect at H^-, then H^- is the center of homothety $\mathcal{H}^-$.

(c) If ω_1 and ω_2 are internally tangent at T, then T is the center of $\mathcal{H}^+$. If they are tangent at T externally, then T is the center of $\mathcal{H}^-$.

Proof. (a) Let AB and $A'B'$ be parallel diameters of ω_1, ω_2, respectively.

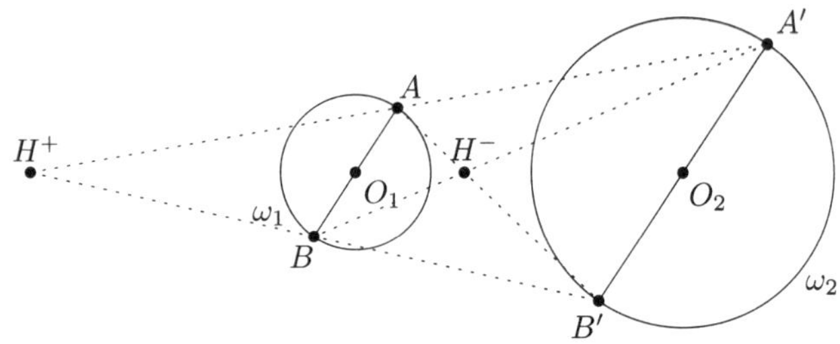

By Proposition 1.28(a) there exists unique homothety that maps A to A' and B to B' and unique homothety that maps A to B' and B to A'. Both such homotheties map ω_1 to a circle with center O_2 and radius O_2A' which is precisely ω_2.

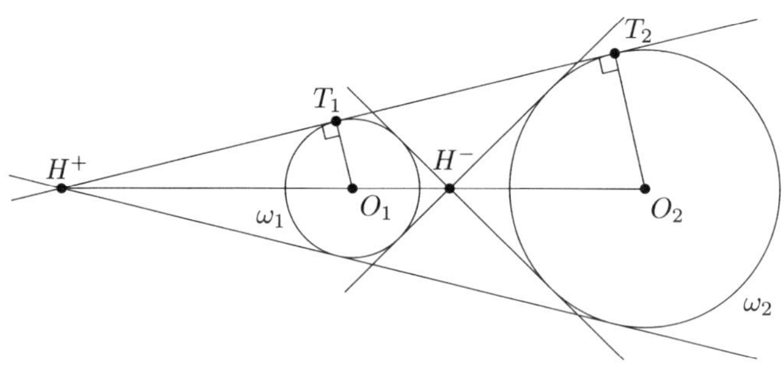

(b) It suffices to prove that H^+ lies on the line O_1O_2 and that $\frac{H^+O_2}{H^+O_1} = \frac{r_2}{r_1}$. The former is clear from symmetry, the latter follows once we denote by T_1, T_2 the points of tangency of one common external tangent and circles ω_1, ω_2, respectively. Then $\triangle H^+O_1T_1 \sim \triangle H^+O_2T_2$ (AA) and hence

$$\frac{H^+O_2}{H^+O_1} = \frac{T_2O_2}{T_1O_1} = \frac{r_2}{r_1}.$$

The part concerning H^- is done similarly.

(c) Finally, if the circles are tangent at T, it is sufficient to prove that $\frac{TO_2}{TO_1} = \frac{r_2}{r_1}$ but this is obvious since $TO_2 = r_2$ and $TO_1 = r_1$. □

Example 1.4. *Circles ω_1, ω_2 are internally tangent at T. Chord AB of ω_1 is tangent to ω_2 at D. Show that TD bisects the angle ATB.*

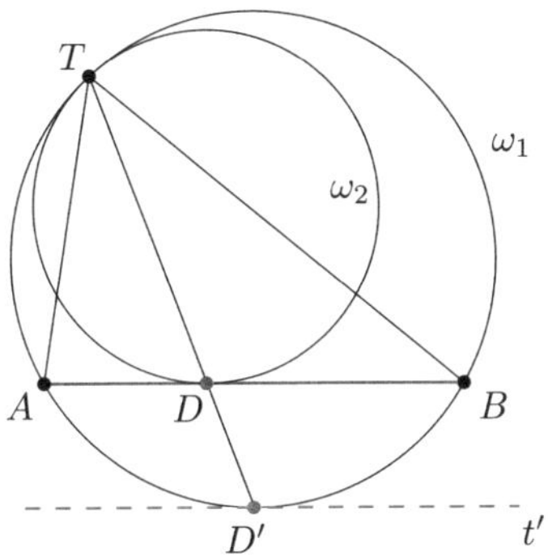

Proof. Extend TD to meet ω_1 for the second time at D'. Since T is the center of a homothety which maps ω_2 to ω_1, point D' is the image of D and the tangent t' to ω_1 at D' is parallel to AB (the tangent to ω_2 at D). This means that D' is the midpoint of arc AB not containing T. The arcs AD' and $D'B$ are then equal and so are the corresponding inscribed angles $\angle ATD'$ and $\angle D'TB$. □

Example 1.5. *Let t be a line. Circles ω_1, ω_2, both lying on the same side of t, are tangent to it at T, U, respectively. Circle ω does not intersect t and is externally tangent to ω_1, ω_2 at K, L, respectively. Show that TK, UL, and ω pass through a common point.*

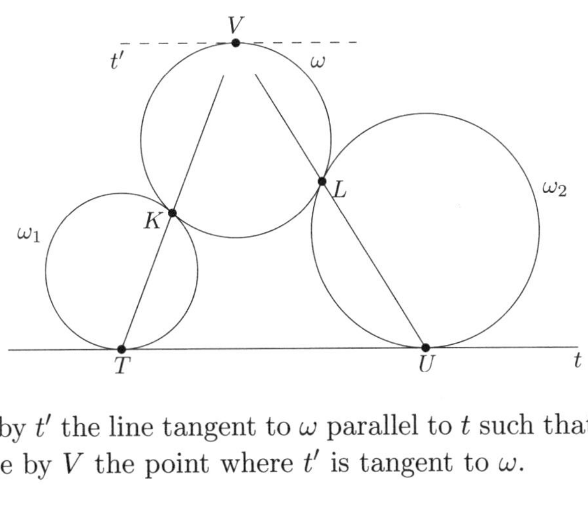

Proof. Denote by t' the line tangent to ω parallel to t such that ω lies between t and t'. Denote by V the point where t' is tangent to ω.

Homothety $\mathcal{H}_1$ with center K that maps ω_1 to ω sends t to t' and hence T to V implying that points T, K and V are collinear. Analogously, homothety $\mathcal{H}_2$ with center L that maps ω_2 to ω sends t to t' and thus U to V, so U, L, V are also collinear and we are done. □

The previous example is rather apparent if one without loss of generality places line t horizontally with ω_1, ω_2 "above" it. The argument then in fact states that homothety with negative factor sends points from the "bottom" to the "top" and vice versa. With this notion the following proposition is immediate!

Proposition 1.30. *Let ABC be a triangle and let its incircle ω and the A-excircle ω_a touch the side BC at D, E, respectively. Let K be the point on the incircle such that KD is a diameter. Then A, K, E lie on a single line.*

Proof. We place BC horizontally with A "above" it.

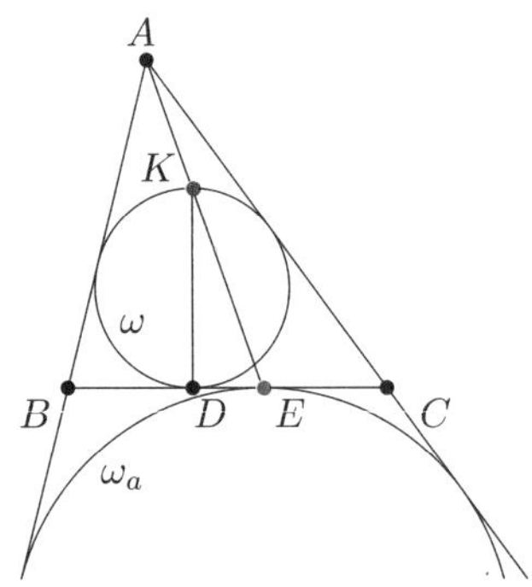

Then E is the "top" point on ω_a and K, as it is antipodal to D, is the "top" point on ω. Thus, these points correspond in the positive homothety which takes ω to ω_a. Since this homothety has center in A (see Proposition 1.29), the points A, K, E are collinear. □

The following two examples are a bit more challenging.

Example 1.6 (IMO 2005 shortlist). *In a triangle ABC satisfying $AC+BC = 3 \cdot AB$ the incircle has center I and touches the sides BC and CA at D and E, respectively. Let K and L be the reflections of the points D and E with respect to I. Prove that the points A, B, K, L lie on one circle.*

Proof. Using Proposition 1.7 (a), the condition can be rewritten as $AB = \frac{1}{2}(AC + BC - AB) = DC = EC$.

Let D' be the point of contact of the A-excircle with side BC. By Proposition 1.7 (c) we have $BD' = DC$, so triangle ABD' is isosceles and $AD' \perp BI$.

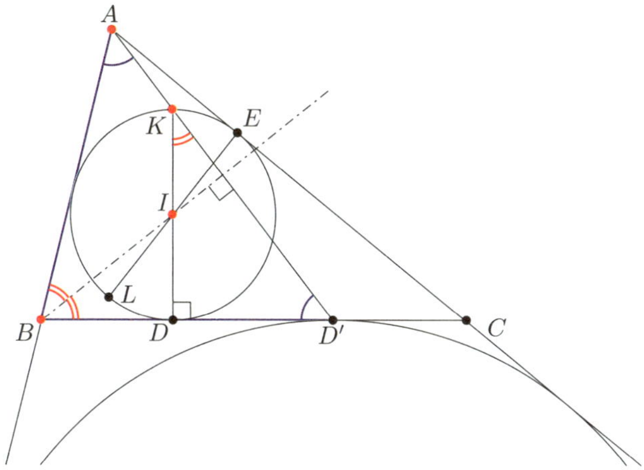

Moreover, points A, K, D' are collinear (see Proposition 1.30). Hence by simple angle-chasing

$$\angle DKD' = 90^\circ - \angle KD'B = \angle D'BI = \angle IBA,$$

and the quadrilateral $ABIK$ is cyclic. Similarly, quadrilateral $ABLI$ is cyclic, so the points A, B, K, L lie on one circle. □

Example 1.7 (USA TST 2010). *Let ABC be a triangle. Points M and N lie on the sides AC and BC, respectively, such that $MN \parallel AB$. Points P and Q lie on the sides AB and CB, respectively, such that $PQ \parallel AC$. The incircle of triangle CMN touches segment AC at E. The incircle of triangle BPQ touches segment AB at F. Lines EN and AB meet at R, and lines FQ and AC meet at S. Given that $AE = AF$, prove that the incenter of triangle AEF lies on the incircle of triangle ARS.*

Proof. Let BC be horizontal.

Since $AE = AF$, there exists a circle ω tangent to AB, AC at F, E, respectively. We claim that ω is in fact the incircle of triangle ARS. Denote by F_1, E_1 the "bottom" points of the incircles of triangles BPQ and CMN, respectively, and by D the "bottom" point of ω.

Consider homothety $\mathcal{H}$ centered at F that maps the incircle of triangle BPQ to ω. Clearly, $\mathcal{H}$ sends segment PQ to AS and point F_1 to D. Thus, it sends segment F_1Q to DS implying that DS is tangent to ω. Similarly, we get that RD is tangent to ω, so ω is indeed the incircle of ARS.

The rest is just some angle-chasing. Focus on triangle ARS, denote by I its incenter and let J be the intersection of ω and segment AI. We want to prove that J is the incenter of triangle AEF.

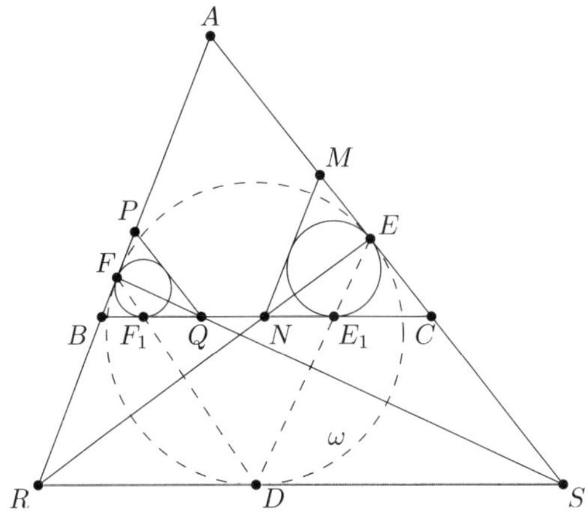

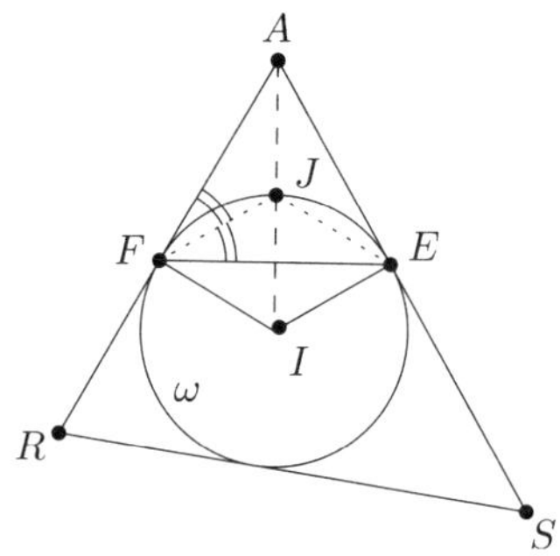

One of the possible approaches is to realize that by symmetry, AJ bisects $\angle FAE$ and that $JF = JE$. Then $\angle EFJ = \angle JEF = \angle JFA$, where the second equality follows from tangency and thus also FJ bisects $\angle EFA$. □

Multiple homothety

After making ourselves well acquainted with homothety, it is time to discuss what happens if we perform two homotheties one after the other.

If these homotheties share the center, the result is obviously a homothety with the same center. If they don't, the question is more interesting. It turns out that (usually) the result is again a homothety. Moreover, the center of this homothety is restricted to lie on the line through the centers of the "partial" homotheties. This is the content of the following lemma which we utilize extensively for the rest of this section.

Lemma 1.31. *Let $\mathcal{H}_1(H_1, k_1)$, $\mathcal{H}_2(H_2, k_2)$ be homotheties such that $H_1 \neq H_2$ and $k_1k_2 \neq 1$. Then their composition (i.e. the transformation of the plane in which we perform $\mathcal{H}_1$ first and then apply $\mathcal{H}_2$ to the result) is again a homothety with center on the line H_1H_2.*

Proof. Once we know what to prove, it is no longer hard. Let AB be a fixed segment and suppose that $\mathcal{H}_1$ maps it to the segment $A'B'$ which in turn is by $\mathcal{H}_2$ mapped to $A''B''$.

Since both $\mathcal{H}_2$ and $\mathcal{H}_1$ are homotheties we have

$$A''B'' \parallel A'B' \parallel AB \quad \text{and} \quad A''B'' = k_2 \cdot A'B' = (k_1k_2) \cdot AB.$$

As $k_1k_2 \neq 1$, the segments AB and $A''B''$ are parallel and of different length, hence there exists a homothety $\mathcal{H}(H,k)$ which maps AB to $A''B''$ (see Proposition 1.28 (a)).

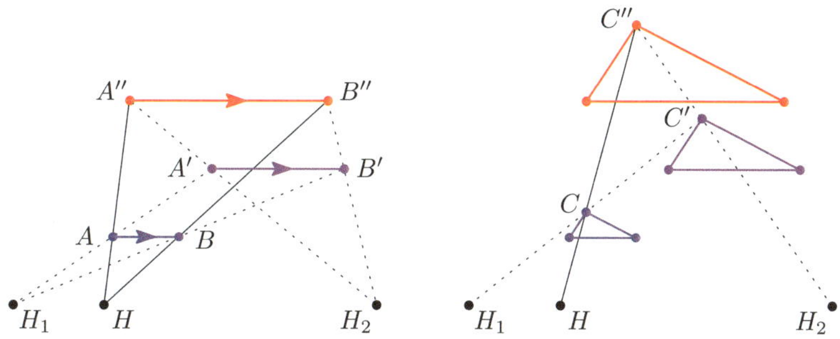

Next we argue that $\mathcal{H}$ in fact works for every point in the plane. Indeed, let C be an arbitrary point, C' its image in $\mathcal{H}_1$, and C'' the image of C' in $\mathcal{H}_2$. Then triangles ABC, $A'B'C'$, and $A''B''C''$ are mutually similar and have corresponding sides parallel, so $\mathcal{H}$ maps not only AB to $A''B''$ but also C to C''. Therefore, the composition of $\mathcal{H}_1$ and $\mathcal{H}_2$ is the homothety $\mathcal{H}$.

Regarding the center of $\mathcal{H}$, observe that H_1 is fixed in $\mathcal{H}_1$ and its image in $\mathcal{H}_2$ belongs to the line H_1H_2. Hence the center of $\mathcal{H}$ lies on the line H_1H_2 which finishes the proof of the lemma.

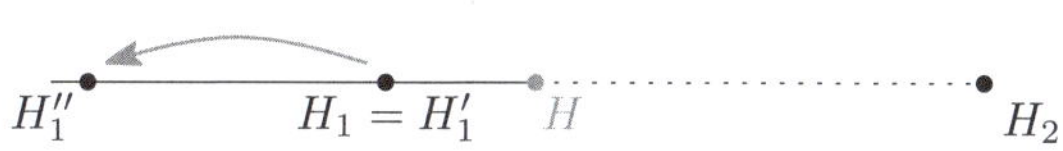

□

The reader is encouraged to verify that (in the setting of the lemma) if $k_1k_2 = 1$ then performing homotheties $\mathcal{H}_1$, $\mathcal{H}_2$ results in translation along the line H_1H_2.

Also, it is worth emphasizing that the resulting homothety has negative factor if and only if exactly one of the "partial" homotheties has negative factor.

Next we introduce one direct corollary of the lemma, namely a stunning theorem of Monge[5].

[5]Gaspard Monge (1746–1818) was a French mathematician who is nowadays considered the "father of descriptive geometry".

Theorem 1.32 (Monge's Theorem). *Let ω_1, ω_2, ω_3 be circles such that the common external tangents of ω_1 and ω_2 intersect at H_3, those of ω_2 and ω_3 intersect at H_1, and those of ω_3 and ω_1 intersect at H_2. Then the points H_1, H_2, H_3 are collinear.*

Proof. Observe that H_3, H_1, H_2 are the centers of positive homotheties which map ω_1 to ω_2, ω_2 to ω_3, and ω_1 to ω_3, respectively. Since the third homothety is the composition of the first two (in other words, ω_1 can be scaled to ω_3 either "directly" or "via" ω_2), its center H_2 lies on the line H_1H_3.

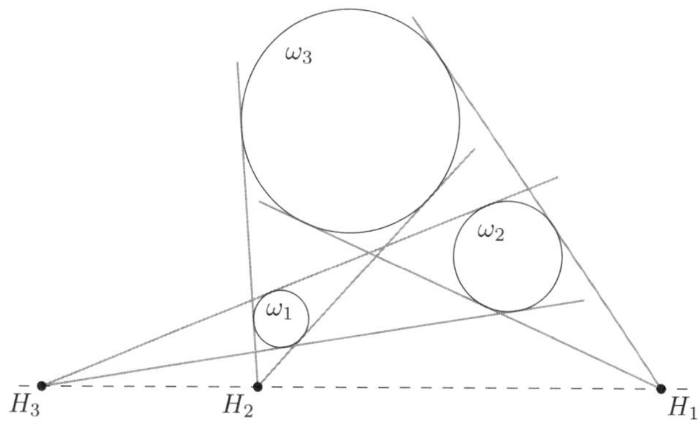

□

The lemma is not only of theoretical significance, but it also provides us with a very efficient way to prove concurrence, in particular when the problem involves tangent circles. The trick is to prove that the desired point of concurrence is in fact the center of some homothety. Just like in the following example which deals with so-called *mixtilinear incircles.*

Example 1.8. *Let ABC be a triangle inscribed in circle Ω. Denote by ω_a the circle tangent to its sides AB, AC and moreover internally tangent to Ω at A_1. Define ω_b, B_1, ω_c, and C_1 analogously. Prove that lines AA_1, BB_1, CC_1, and OI are concurrent, where O and I denote the circumcenter and the incenter of triangle ABC, respectively.*

Proof. Observe that A_1 is the center of positive homothety between Ω and ω_a (if in doubt, consult Proposition 1.29) and that A is the center of positive homothety between ω_a and the incircle ω of triangle ABC (sides AB, AC are their common external tangents). Hence line AA_1 passes through the center H^+ of positive homothety between ω and Ω.

Likewise, BB_1 and CC_1 pass through H^+. Finally, as the homothety which maps ω to Ω also maps I to O, point H^+ belongs to OI too. The concurrence is thus established. □

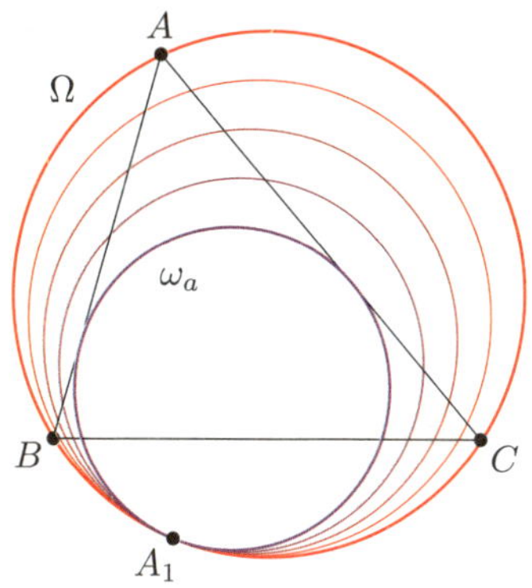

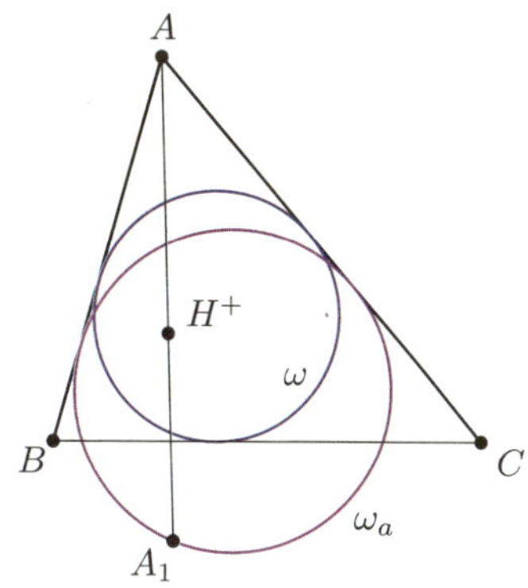

We end this section with one slightly less straightforward example.

Example 1.9. *Points K, L, M, N lie on the sides AB, BC, CD, DA of a quadrilateral $ABCD$, respectively, such that lines AB, CD, and LN are concurrent at P and lines AD, BC, and KM are concurrent at Q. Denote by X the intersection of KM and LN. Prove that if the quadrilaterals $AKXN$, $BLXK$, and $CMXL$ have inscribed circles then the quadrilateral $DNXM$ has one too.*

Proof. We aim to make use of the Lemma 1.31 again. Denote the circles inscribed in quadrilaterals $AKXN$, $BLXK$, and $CMXL$ by ω_a, ω_b, ω_c, respectively. Further, let ω_d be the circle tangent to segment XM and rays XN and MD. We aim to prove that ω_d is actually tangent to DN too.

First we map ω_a to ω_c via ω_b. Since P is the center of positive homothety between ω_a and ω_b and Q is the center of positive homothety between ω_b and ω_c, the center of positive homothety between ω_a and ω_c (call it H) belongs to the line PQ.

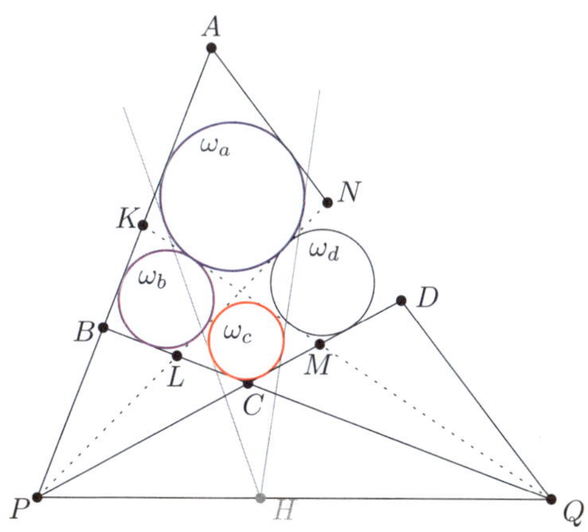

Next we map ω_a to ω_d via ω_c. As above we realize that the center of positive homothety between ω_a and ω_d lies on the line HP which coincides with PQ. However, this center also has to lie on the common external tangent QK of ω_a and ω_d, hence the center of positive homothety between ω_a and ω_d is in fact Q.

Finally, since ω_a is tangent to QA, so is its image ω_d in homothety with center Q. $\square$

Exploring the Triangle

The most important point of focus in Euclidean geometry is certainly the geometry of a triangle. It has been investigated for thousands of years and new results are still produced. As of 2018, over fifteen thousand interesting points have been located in a triangle! For the purposes of this book, we will concentrate on the two most frequent configurations. Namely those containing the orthocenter and the incenter.

Orthocenter, nine-point circle

We will see that the orthocenter is in some sense the most convenient point in a triangle. The main reason is that due to right angles, many circles are involved, and thus angle-chasing is (with few exceptions) a sure-fire strategy.

Proposition 1.33. *Let ABC be a triangle with orthocenter H. Then H lies inside the triangle if and only if the triangle is acute.*

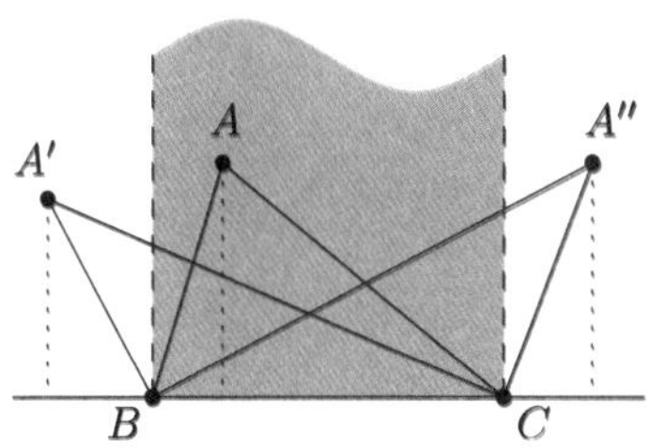

Proof. Since H lies on the altitude from vertex A, we may observe that it lies inside the half-strip erected on BC if and only if both angles $\angle B$, $\angle C$ are acute. By applying an analogous argument we obtain that H lies inside ABC (inside half-strips over all three sides) if and only if all angles in ABC are acute. □

Note that in a right triangle the orthocenter coincides with the vertex opposite to the hypotenuse and the picture degenerates. For this reason we will exclude right triangles from further considerations in this section.

The following lemma is extremely useful when discussing the case of an obtuse triangle in problems where the orthocenter is present. It basically says that we are still dealing with the same picture.

Lemma 1.34. *Let ABC be a triangle with orthocenter H. Then the orthocenters of triangles BCH, CAH, ABH are points A, B, C, respectively.*

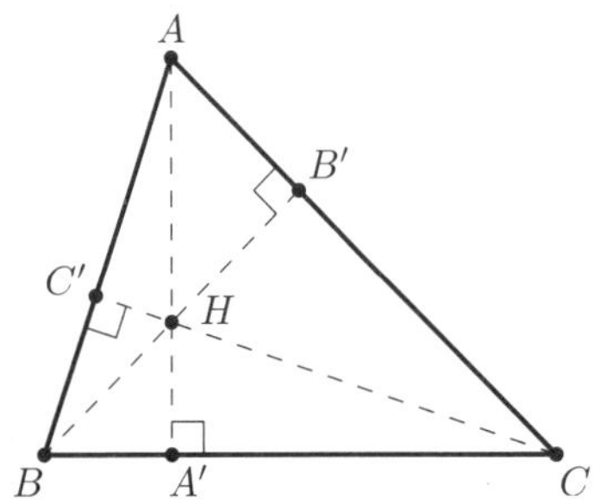

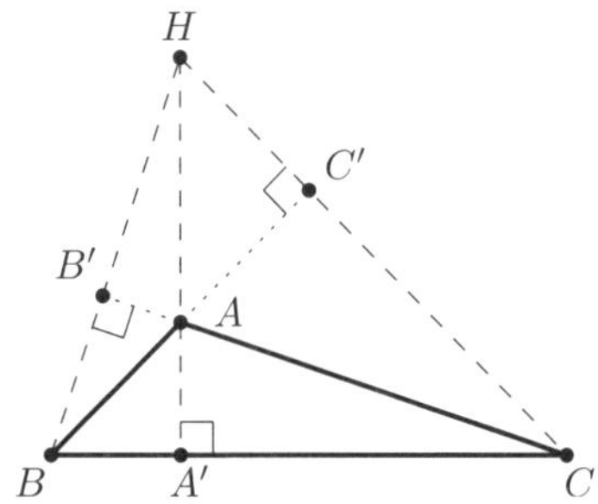

Proof. Lines AH, AB, AC are in fact altitudes in triangle HBC, because $AH \perp BC$, $AB \perp CH$, and $AC \perp HB$. Hence A is the orthocenter in triangle HBC. The rest follows from an analogous argument. □

Proposition 1.35 (Basic properties of the orthocenter). *Let AA', BB', CC' be the altitudes in triangle ABC with orthocenter H and circumradius R. Then:*

(a) Quadrilaterals[6] $BCB'C'$, $CAC'A'$, $ABA'B'$ are cyclic with sides BC, CA, AB, respectively, as their diameters.
(b) Quadrilaterals[6] $AC'HB'$, $BA'HC'$, $CB'HA'$ are cyclic with segments AH, BH, CH, respectively, as diameters.
(c) If angles $\angle B$ and $\angle C$ are acute, then $\angle BHC = 180° - \angle A$ and otherwise $\angle BHC = \angle A$.
(d) The circumradii of triangles BHC, CHA, AHB are all equal to R.
(e) Triangles $AB'C'$, $A'BC'$, $A'B'C$ are all similar to triangle ABC with ratios of similitude equal to $|\cos \angle A|$, $|\cos \angle B|$, $|\cos \angle C|$, respectively.
(f) $AH = 2R|\cos \angle A|$, $BH = 2R|\cos \angle B|$, $CH = 2R|\cos \angle C|$.

Proof. In (a), quadrilateral $BCB'C'$ is cyclic with diameter BC since $\angle BB'C = 90° = \angle CC'B$. For the others the situation is analogous.

Similarly in part (b), $AC'HB'$ is inscribed in a circle with diameter AH as $\angle AC'H = 90° = \angle HB'A$. The rest follows by analogy.

For (c), we use the circle through A, B', H, and C'. With the help of the previous proposition we infer that both $\angle B$ and $\angle C$ are acute if and only if angles $\angle A$ and $\angle C'HB'$ $(= \angle BHC)$ intercept the chord $B'C'$ from opposite half-planes. In either case we obtain the conclusion.

In (d), we write the result of (c) as $\sin \angle BHC = \sin \angle A$ regardless of whether triangle ABC is acute.

Then by the Extended Law of Sines the circumradius R_1 of triangle BHC equals

$$R_1 = \frac{BC}{2\sin \angle BHC} = \frac{BC}{2\sin \angle A} = R.$$

[6] Possibly in different order of vertices.

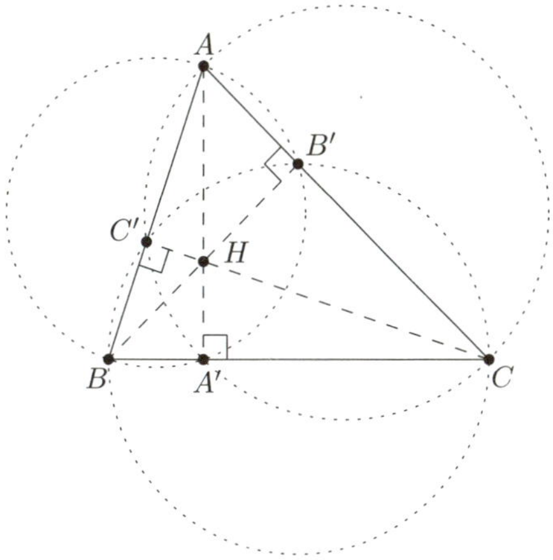

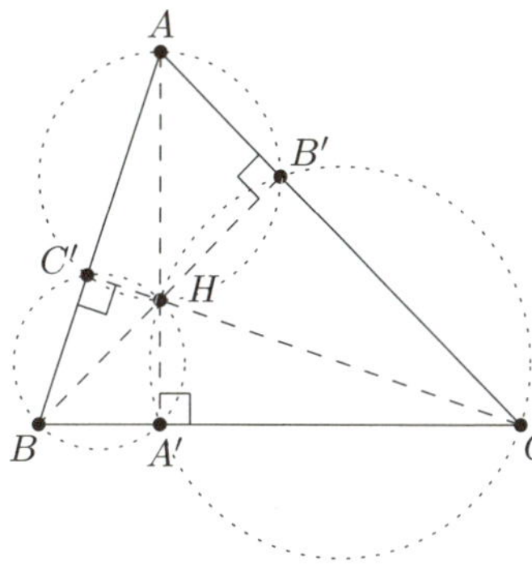

In (e), the similarity $\triangle AB'C' \sim \triangle ABC$ follows from the concyclicity of $BCB'C'$. The ratio of similitude equals $\frac{AC'}{AC}$, which from the right triangle ACC' is equal to $|\cos\angle A|$.

For part (f), since by (b) AH is a diameter of the circumcircle of triangle $AB'C'$ and the diameter of the circumcircle of triangle ABC is $2R$, we can conclude by (e). □

There is still more to come!

Proposition 1.36 (Reflections of the orthocenter). *Let ABC be a triangle with orthocenter H. Denote by H_a the reflection of H over the side BC and denote by H_a' the image of H under reflection about the midpoint of BC. Define points H_b, H_b', H_c, H_c' analogously. Then points H_a, H_a', H_b, H_b', H_c, H_c' lie on the circumcircle ω of triangle ABC and AH_a', BH_b', CH_c' are its diameters.*

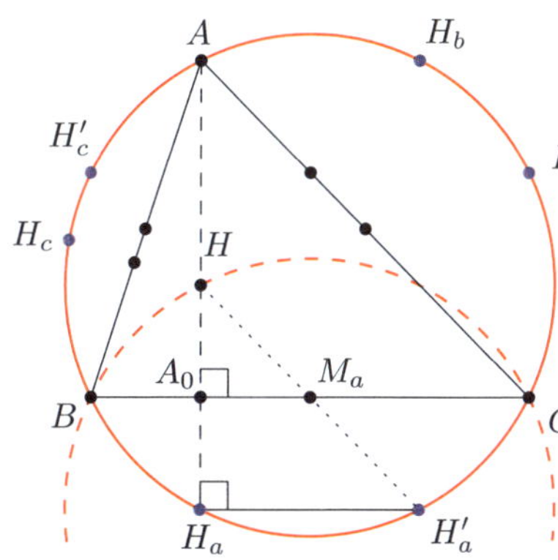

Proof. Since the circumcircles of triangles ABC and BHC have equal radii (see Proposition 1.35(d)), they are in fact symmetric in line BC. Thus H_a

being the symmetric point to H indeed lies on ω. For H'_a we note that the two circumcircles are symmetric also with respect to the midpoint of BC and repeat the same argument.

If $AB = AC$, the last part follows from symmetry. Otherwise triangles $HH_aH'_a$ and HA_0M_a, where M_a and A_0 are the midpoint of BC and the foot of the A-altitude on BC, are homothetic with center H and factor 2. Therefore

$$\angle AH_aH'_a = \angle HH_aH'_a = \angle HA_0M_a = 90^\circ$$

and AH'_a is indeed diameter of ω. □

The most important discovery in this configuration was made by J. V. Poncelet[7] in 1821. It concerns yet another circle.

Theorem 1.37 (The nine-point circle). *Let AA', BB', CC' be the altitudes in triangle ABC with orthocenter H, circumcenter O and circumradius R. Denote by M_a, M_b, M_c the midpoints of the sides BC, CA, AB, respectively, and let N_a, N_b, N_c be the midpoints of the segments AH, BH, CH, respectively. Then points M_a, M_b, M_c, A', B', C', N_a, N_b, N_c lie on a circle with radius $\frac{R}{2}$. The center O_9 of this circle bisects the segment OH. Segments N_aM_a, N_bM_b, N_cM_c are diameters of the circle.*

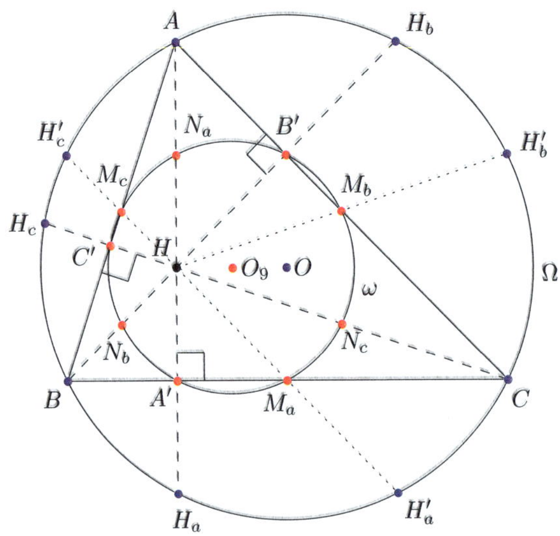

Proof. We just take the configuration from Proposition 1.36 and apply homothety $\mathcal{H}(H, \frac{1}{2})$. The conclusion follows. □

[7]Jean-Victor Poncelet (1788–1867) was a French engineer and mathematician.

We also proved that the center O_9 of the nine-point circle lies on the Euler line of triangle ABC (see Example 1.3).

Next we show a typical angle-chasing problem involving the orthocenter.

Example 1.10. *Let AK, BL, CM be the altitudes of an acute triangle ABC and H its orthocenter. Let $S = BL \cap KM$, P the midpoint of AH and $T = LP \cap AM$. Show that $TS \perp BC$.*

Proof. It suffices to show that $TS \parallel AK$ or in other words $\angle MTS = \angle BAK$. But since $\angle BAK = \angle MAH = \angle MLH$ as $MHLA$ is cyclic (see Proposition 1.35(b)) we in fact need $\angle MTS = \angle MLS$ or the quadrilateral $TMSL$ to be cyclic.

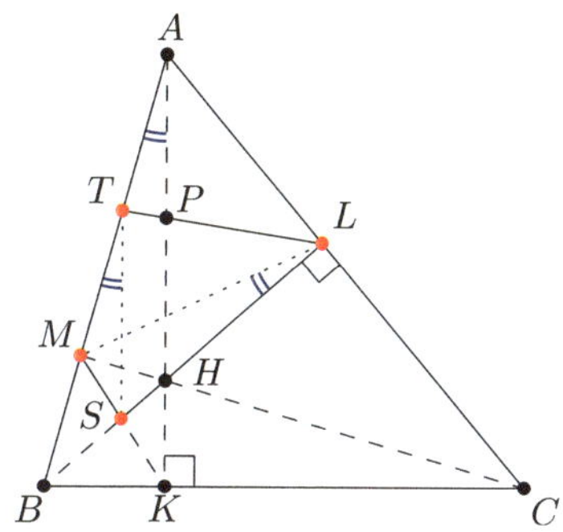

This should not be difficult as after a quick glance we see that angles SMT and TLS can be expressed in terms of $\angle A$, $\angle B$, $\angle C$. Indeed, since $KCAM$ is cyclic $\angle SMT = 180° - \angle C$.

For $\angle TLS$ we first calculate $\angle ALP$. Knowing that triangle ALP is isosceles (PA and PL are radii of the circumcircle of $MHLA$) we may write $\angle ALP = \angle PAL = 90° - \angle C$. Thus $\angle TLS = 90° - \angle ALP = \angle C$.

We obtained $\angle SMT + \angle TLS = 180°$, thus $TMSL$ is cyclic and the proof is complete. □

Example 1.11 (All-Russian Olympiad 2008). *In an acute triangle ABC the altitudes BB_1 and CC_1 intersect at H, O is the circumcenter, and M the midpoint of the side BC. The line AO intersects the side BC at P, while the lines AH and B_1C_1 meet at Q. Prove that the lines HM and PQ are parallel.*

Proof. Draw the circumcircle ω of triangle ABC and let H'_a be the image of H under reflection about M. Then H, M, H'_a are collinear and also A, O, H'_a are collinear as AH'_a is a diameter of ω (see Proposition 1.36).

In order to prove $HH'_a \parallel PQ$ it suffices to prove that triangles AQP and AHH'_a are similar. Since these triangles share one angle, we need $\frac{AQ}{AP} = \frac{AH}{AH'_a}$. By Propositions 1.35(f) and 1.36 we have

$$\frac{AH}{AH'_a} = \frac{2R\cos\angle A}{2R} = \cos\angle A.$$

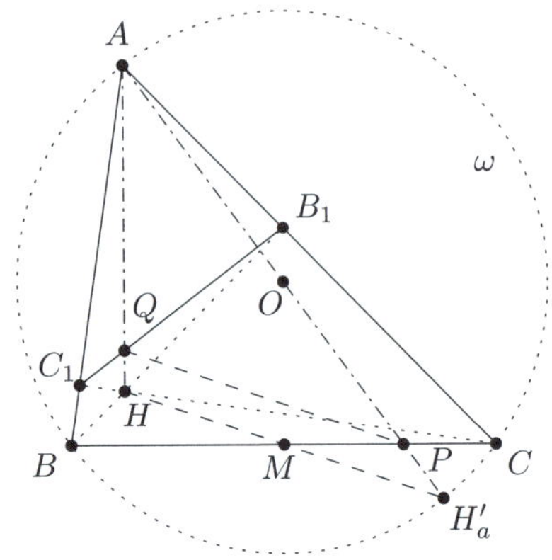

On the other hand, segments AQ, AP are corresponding cevians (both pass through the respective circumcenters) in similar triangles AB_1C_1, ABC, so from Proposition 1.35(e) we also obtain that $\frac{AQ}{AP} = \cos\angle A$. Hence the triangles AQP and AHH'_a are similar and the conclusion follows. □

Sometimes it is important to realize that what we were given in a problem is some part of a well-known configuration. Restoring the rest of it is often the winning strategy. Like in the next example.

Example 1.12 (China Western MO 2010). *Quadrilateral $ABCD$ is inscribed in a semicircle with diameter AB and center O. Lines tangent to the semicircle at points C and D meet at E and the segments AC and BD meet at F. Denote by M the intersection of EF and AB. Prove that E, C, M, and D are concyclic.*

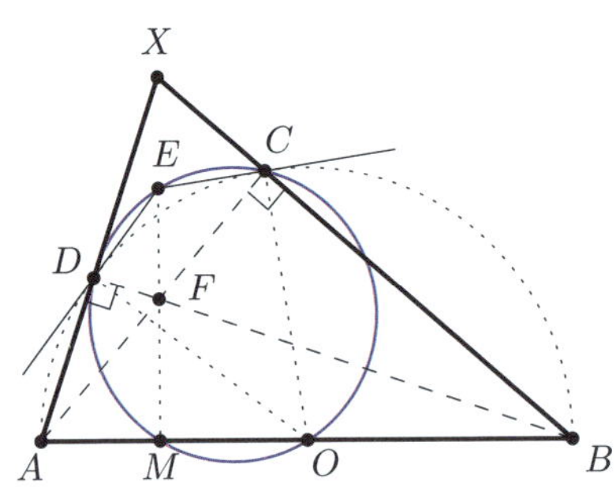

Proof. Let AD and BC intersect at X. Now we recognize that F is the orthocenter in triangle ABX. Points O, C, D lie on the nine-point circle of triangle ABX and $\angle ODE = \angle OCE = 90°$, so E must be antipodal point to O on the nine-point circle. Thus E is the midpoint of FX implying that M is the foot of the altitude from X. As such it also lies on the nine-point circle of triangle ABX. □

Incenter, Midpoint of Arc

The second point we shall discuss is the incenter. Quite surprisingly, despite its close relation to the incircle, its fundamental properties are more related to the circumcircle of a triangle. This is due to the fact that angle bisectors have nice angular properties. In particular, they bring the midpoints of arcs into play.

Proposition 1.38 (Basic properties of the incenter). *In triangle ABC inscribed in a circle ω let I be the incenter, M the midpoint of arc BC of ω that does not contain A, and D the foot of the A-angle bisector. Then:*

(a) $\angle BIC = 90^\circ + \frac{1}{2}\angle A$.
(b) *M lies on the angle bisector of $\angle A$ and $MB = MC = MI$.*
(c)

$$\frac{AI}{ID} = \frac{b+c}{a}.$$

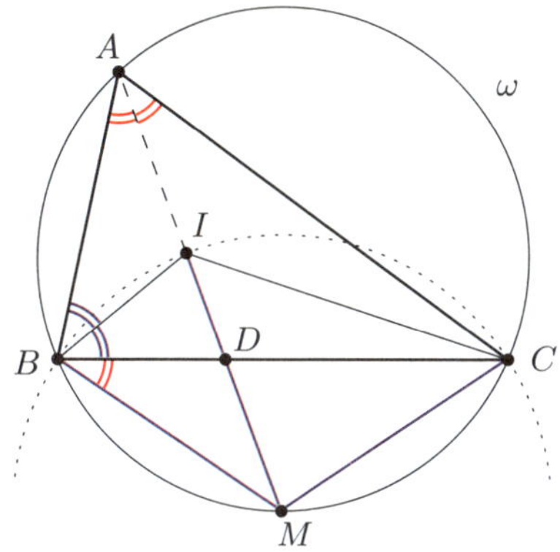

Proof. For (a), in triangle BIC we have $\angle BIC = 180^\circ - \frac{1}{2}\angle B - \frac{1}{2}\angle C = 90^\circ + \frac{1}{2}\angle A$.

In part (b), the arcs MB and MC are equal, hence the corresponding inscribed angles are also equal and AM bisects the angle BAC. The equality $MB = MC$ is obvious and it suffices to show that triangle IBM is isosceles. Luckily, that's simple:

$$\angle BIM = 180^\circ - \angle AIB = \frac{1}{2}\angle A + \frac{1}{2}\angle B$$

and

$$\angle MBI = \angle MBC + \angle CBI = \frac{1}{2}\angle A + \frac{1}{2}\angle B.$$

We infer $MI = MB$ and conclude the proof of part (b).

Finally in (c) we apply the Angle Bisector Theorem in triangles ABD and ABC to learn the desired

$$\frac{AI}{ID} = \frac{AB}{BD} = \frac{c}{\frac{ac}{b+c}} = \frac{b+c}{a}.$$

□

Example 1.13 (IMO 2006). *Let ABC be a triangle with incenter I. A point P in the interior of the triangle satisfies*

$$\angle PBA + \angle PCA = \angle PBC + \angle PCB.$$

Show that $AP \geq AI$, and that equality holds if and only if $P = I$.

Proof. First, we analyze the condition.

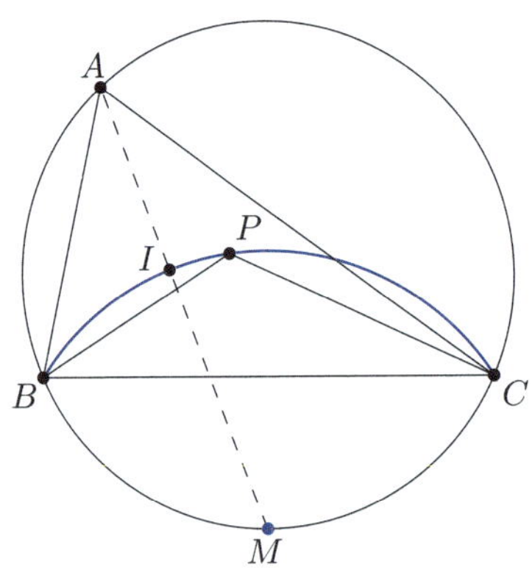

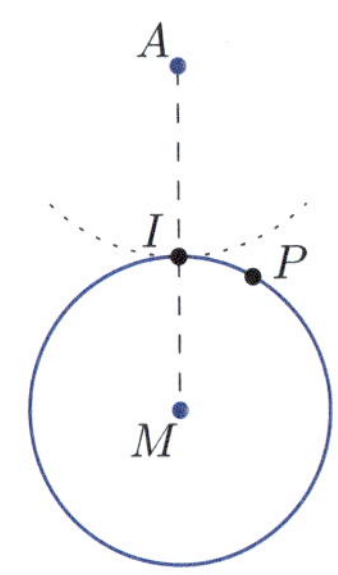

Since the sum of both its sides equals $\angle B + \angle C$, simple angle-chasing gives us

$$\angle BPC = 180^\circ - (\angle PBC + \angle PCB) = 180^\circ - \frac{1}{2}(\angle B + \angle C) = 90^\circ + \frac{1}{2}\angle A.$$

Thus by Proposition 1.38(a) point P lies on the arc BIC.

Now the key is to recall that the circumcenter of triangle BIC is the midpoint M of arc BC that does not contain A. In particular, it is a point on the line AI. Now the conclusion follows just by looking at the picture! Indeed, among all the points on the circumcircle of triangle BIC, point I is the one closest to A.

(The rigor-seeking reader may for $P \neq I$ write down the triangle inequality in triangle AMP and subtract $MI = MP$.) □

Now we will form alternative definitions of the incenter of a triangle. They are often useful, especially in problems, where only one angle bisector is involved.

Proposition 1.39 (Alternative definitions of the incenter). *In triangle ABC let I be the incenter, M the midpoint of arc BC that does not contain A, and let $D = AI \cap BC$. Let X be a point on segment AD. The following statements are equivalent:*

(a) $X = I$.
(b) $MX = MI$.
(c) $\angle BXC = 90^\circ + \frac{1}{2}\angle A$.

Proof. We already know that I satisfies both (b) and (c) (see Proposition 1.38) so it remains to realize that it is the only point on segment AD with any of these properties.

For (b) it is obvious. For (c), we note that X lies on the circumcircle of triangle BCI which intersects segment AM at one point only. □

Example 1.14 (IMO 2002). *Let BC be a diameter of circle ω centered at O. Let A be a point of ω such that $\angle AOB < 120^\circ$. Let D be the midpoint of the arc AB which does not contain C. The line through O parallel to DA meets the line AC at I. The perpendicular bisector of OA meets ω at E and at F. Prove that I is the incenter of the triangle CEF.*

Proof. Thanks to condition $\angle AOB < 120^\circ$, point A is the midpoint of arc EF which does not contain C. Hence line CA is the angle bisector of $\angle ECF$. It remains to prove $AI = AF$. We claim that both lengths are in fact equal to the radius of the circle ω.

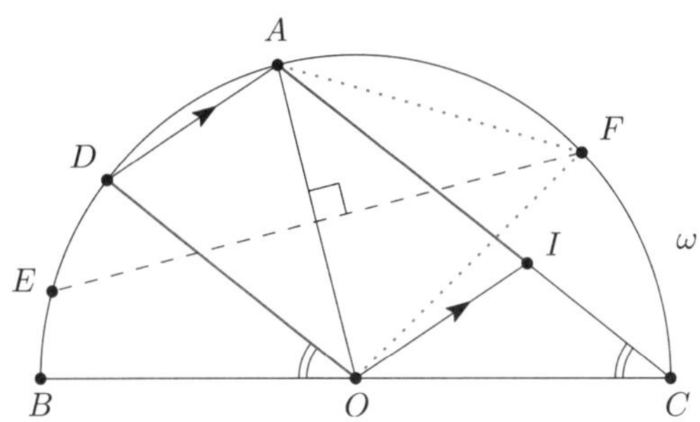

This assertion is obvious for AF because as F lies on the perpendicular bisector of AO, we have $AF = OF$.

Moreover, since D is the midpoint of arc AB, we have $\angle BOD = \frac{1}{2}\angle BOA = \angle BCA$, so $OD \parallel CA$. But this means that quadrilateral $DOIA$ is a parallelogram ($DA \parallel OI$ was given). Thus $AI = DO$ and we are done. □

The points on the angle bisector are tied by many relations. One of them is a consequence of a metric identity which holds in a more general framework. For reference purposes we shall call it the Shooting Lemma.

Proposition 1.40 (Shooting Lemma). *Let M be the midpoint of arc BC of the circle ω. Let ray ℓ emanating from M intersect segment BC at D and ω for the second time at A. Then:*

(a) $MD \cdot MA = MB^2$.
(b) If I is the incenter of triangle ABC, then $MD \cdot MA = MI^2$
(c) If another ray ℓ' from M intersects BC at D' and ω at A', then $DD'A'A$ is cyclic.

Proof. We start with (a). As M is the midpoint of arc BC, we have $\angle MBC = \frac{1}{2}\angle A = \angle MAB$. Hence the line MB is tangent to the circumcircle of triangle ABD (see Proposition 1.15) and by Power of a Point $MD \cdot MA = MB^2$.

Part (b) follows immediately from $MB = MI$ (see Proposition 1.38(b)).

And in part (c) the concyclicity of $DD'A'A$ is ensured by Power of a Point as (a) gives

$$MD \cdot MA = MB^2 = MD' \cdot MA'.$$

□

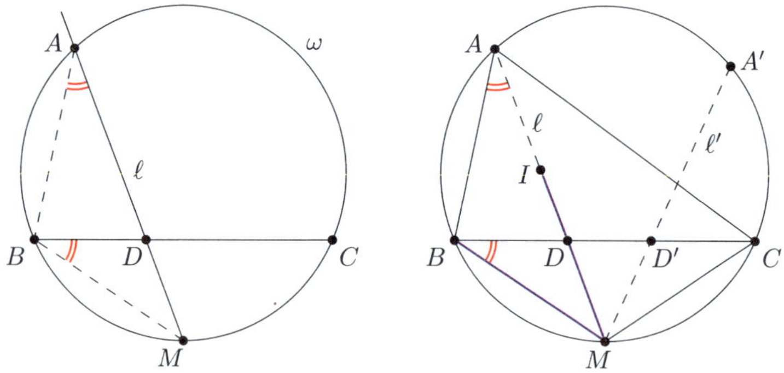

The following proposition reveals strong connections between the incenter and the orthocenter.

Proposition 1.41. *Let $A'B'C'$ be a triangle inscribed in a circle ω and with incenter I. Let M_a, M_b, M_c be the midpoints of arcs $B'C'$, $C'A'$, $A'B'$ of ω that do not contain points A', B', C', respectively. Further, denote by M_a', M_b', M_c' the antipodal points on the circumcircle of triangle $A'B'C'$ with respect to M_a, M_b, M_c, respectively. Then we obtain exactly the same configuration as in Proposition 1.36 where points A', B', C', M_a, M_b, M_c, M_a', M_b', M_c', I correspond to H_a, H_b, H_c, A, B, C, H_a', H_b', H_c', H, respectively (in the notation of Proposition 1.36).*

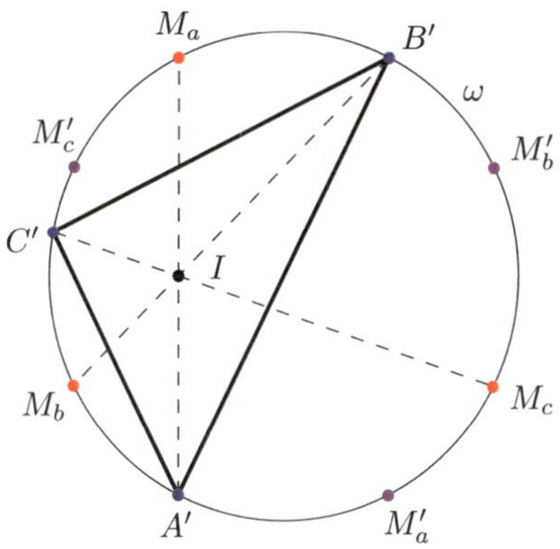

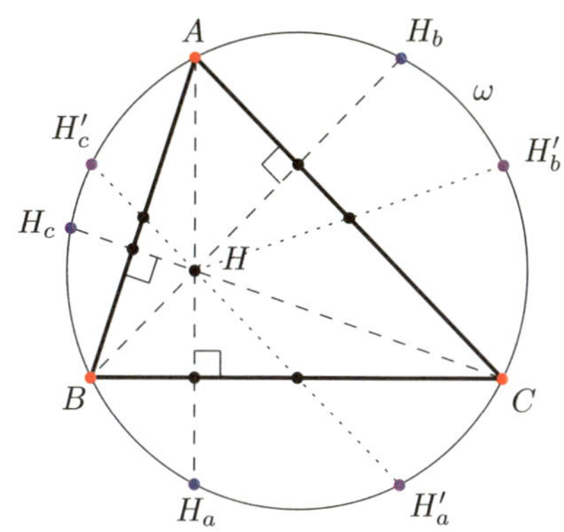

Proof. The angle between lines $A'M_a$ and M_bM_c equals by Corollary 1.14(a) the sum of angles corresponding to the shorter arcs $A'M_c$ and M_bM_a, thus it equals $\frac{1}{2}\angle C' + (\frac{1}{2}\angle A' + \frac{1}{2}\angle B') = 90°$. Hence $A'M_a$ is an altitude in triangle $M_aM_bM_c$. Similarly, $B'M_b$ and $C'M_c$ are altitudes and so I is the orthocenter in triangle $M_aM_bM_c$ and the points A', B', C' correspond to the images of orthocenter in reflection over the triangle sides. Since points M'_a, M'_b, and M'_c are antipodal to M_a, M_b, M_c, respectively, they indeed correspond to images of orthocenter in reflections about the midpoints of the sides of triangle $M_aM_bM_c$ (recall AH'_a is a diameter). □

Excenters, the Big Picture

In order to reveal another strong connection between the incenter and the orthocenter, we add some points in our picture, namely the excenters. Again, we may be surprised that the excenters are in a certain way more compatible with the circumcircle than with the actual excircles.

Now, let's disclose the most significant proposition of this section. Quite unexpectedly the picture we obtain turns out to be rather familiar!

Proposition 1.42 (The Big Picture). *In triangle ABC with incenter I let M_a, M_b, M_c be the midpoints of arcs BC, CA, AB that do not contain points A, B, C, respectively. Further, denote by M'_a, M'_b, M'_c the antipodal points on the circumcircle of triangle ABC with respect to M_a, M_b, M_c, respectively. Finally, let I_a, I_b, I_c be the excenters opposite to vertices A, B, C, respectively. Then I is the orthocenter of triangle $I_aI_bI_c$ and the circumcircle of triangle ABC is the nine-point circle of triangle $I_aI_bI_c$. This has the following consequences:*

(a) Points M'_a, M'_b, M'_c are the midpoints of the respective sides in triangle $I_aI_bI_c$.

(b) *Quadrilaterals $BICI_a$, $CIAI_b$, $AIBI_c$ are cyclic with diameters II_a, II_b, II_c, respectively. The centers of the respective circles are M_a, M_b, M_c.*

(c) *Quadrilaterals I_bI_cBC, I_cI_aCA, I_aI_bAB are cyclic with diameters I_bI_c, I_cI_a, I_aI_b, respectively. The centers of the respective circles are M_a', M_b', M_c'.*

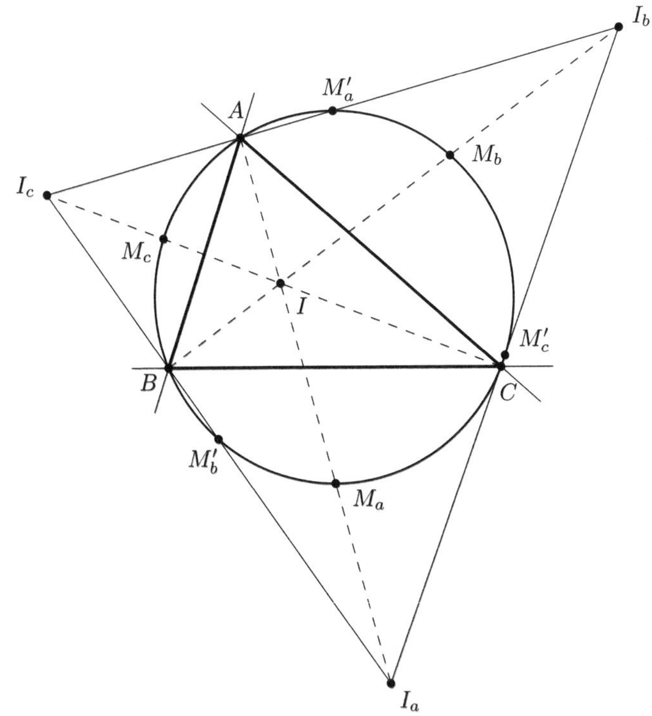

Proof. First observe that points I_b, I_c both lie on the external bisector of $\angle A$ and thus A lies on I_bI_c. Now calculate

$$\angle I_aAI_b = \angle I_aAC + \angle CAI_b = \frac{1}{2}\angle A + \frac{1}{2}(180^\circ - \angle A) = 90^\circ.$$

Hence A is indeed the foot of the altitude in triangle $I_aI_bI_c$. Since an analogous argument holds for B and C, then I is indeed the orthocenter of triangle $I_aI_bI_c$ and thus the circumcircle of triangle ABC is indeed the nine-point circle of triangle $I_aI_bI_c$. □

In the following problems we again apply the idea of integrating the given picture into some well-known configuration.

Example 1.15 (All-Russian Olympiad 2005). *Let ABC be a triangle and I its incenter. Denote by A_1 the midpoint of BC and by M_a' the midpoint of arc BC containing vertex A. Prove that $\angle IA_1B = \angle IM_a'A$.*

Proof. Draw the Big Picture from Proposition 1.42 and observe that since BCI_bI_c is cyclic, the triangles BIC and I_cII_b are similar.

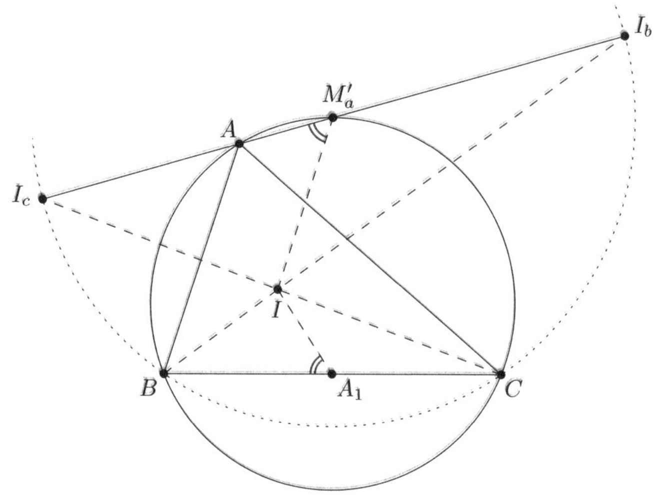

Moreover, IA_1 and IM'_a are corresponding medians in these triangles and angles $\angle IA_1B$ and $\angle IM'_aA$ also correspond in this similarity and thus are equal. □

Example 1.16 (All-Russian Olympiad 2006). *Let ABC be a triangle. The angle bisectors of the angles ABC and BCA intersect the sides CA and AB at points B_1 and C_1, and intersect each other at point I. The line B_1C_1 intersects the circumcircle ω of triangle ABC at points M and N. Prove that the circumradius of triangle MIN is twice as long as the circumradius of triangle ABC.*

Proof. Again, draw the Big Picture! We claim that the circumcircle of triangle MIN is in fact the circumcircle ω_1 of triangle I_bII_c, which we know from Propositions 1.42 and 1.35(d) to have twice as long radius as ω, the nine-point circle of triangle $I_aI_bI_c$.

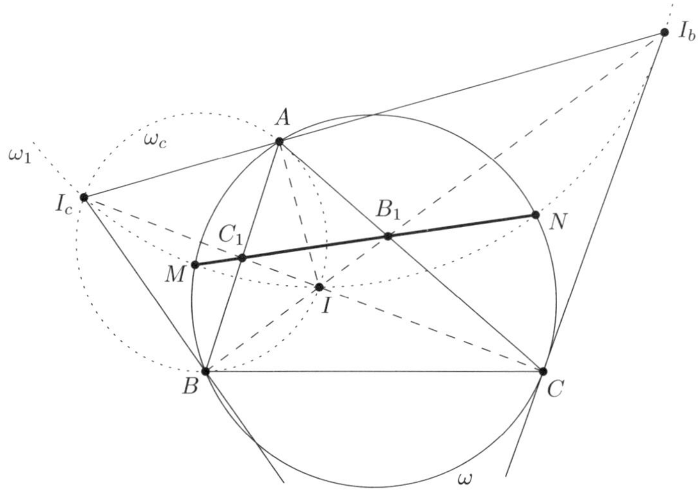

It suffices to prove that B_1 and C_1 lie on the radical axis of ω and ω_1, since then M, N would indeed lie on ω_1.

But this follows from the Radical Lemma as $BIAI_c$ and $CIAI_b$ are cyclic.

$\square$

Spiral Similarity

In the family of geometric transformations there is an exquisite gem with a noble name, the spiral similarity. Mastering this transformation ensures the deepest insight and the techniques we are about to reveal reduce many olympiad problems to simple exercises.

As the name suggests, spiral similarity will also preserve the shape of a figure, but this time also rotation will be involved.

Given a point S, a positive number k, and an angle φ different from $0°$ and $180°$, *spiral similarity* with center S, dilation factor k, and angle of rotation φ is a geometric transformation that sends point A to a point A' such that:

(a) $SA' = k \cdot SA$,
(b) $\angle(SA, SA') = \varphi$.

Such a spiral similarity is denoted by $\mathcal{S}(S, k, \varphi)$. Note that the triangle SAA' will have fixed shape (SAS), regardless of which point A we choose. We can say that this shape is produced by $\mathcal{S}$.

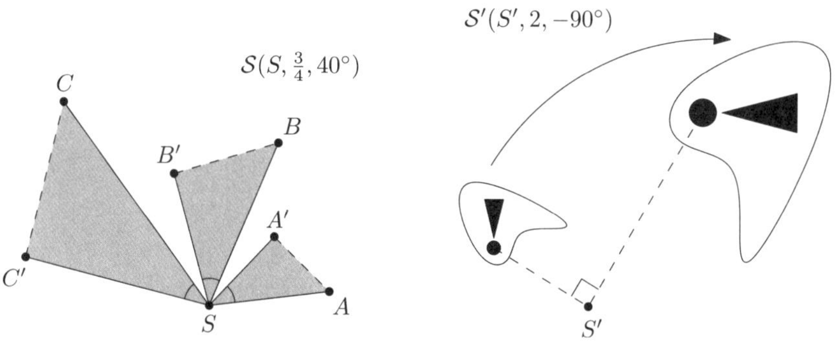

If we allowed $\varphi = 0°$ or $\varphi = 180°$, spiral similarity would reduce to homothety. For $k = 1$ it reduces to rotation. In general, spiral similarity is a composition of these two transformations.

As homothety maps figures to similar figures and spiral similarity is only homothety followed by rotation, it also maps figures to similar figures. Moreover, these two figures are always *directly similar*. This means that the corresponding points of the two figures are arranged in the same (either both in clockwise or both in anti-clockwise) order.

Proposition 1.43. *Let $\mathcal{S}(S, k, \varphi)$ be a spiral similarity. Then:*

(a) Image of a line ℓ is a line. If we denote it by ℓ', then $\angle(\ell, \ell') = \varphi$.
(b) Image of a triangle ABC is a triangle $A'B'C'$ directly similar to it with factor k. In other words,

$$A'B'/AB = A'C'/AC = B'C'/BC = k$$

and

$$\angle(AB, A'B') = \angle(AC, A'C') = \angle(BC, B'C') = \varphi.$$

(c) Image of a circle with radius R is a circle with radius $k \cdot R$.

Proof. We shall prove only (a) and leave (b) and (c) as easy exercises for the reader. The image of line ℓ under homothety $\mathcal{H}(S,k)$ is a line ℓ_2 parallel to ℓ. Image of this line under rotation is again a line.

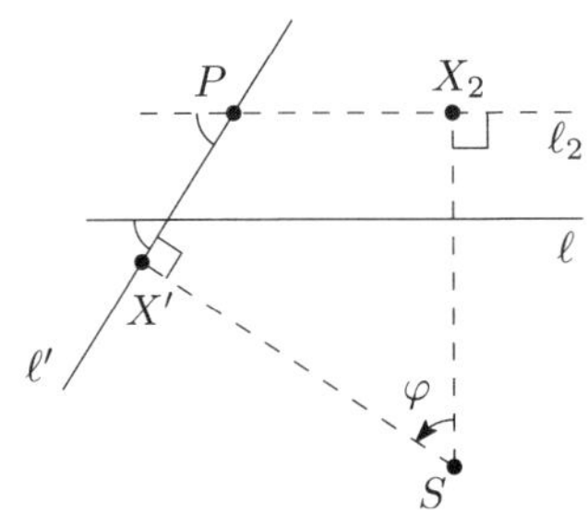

Now denote by X_2 the projection of S onto ℓ_2. Since rotation preserves angles, the image X' of X_2 under the rotation with center S and angle φ is the projection of S onto ℓ'. Thus if we denote the intersection of ℓ_2 and ℓ' by P, we obtain $\angle(\ell, \ell') = \angle(\ell_2, \ell') = \angle(PX_2, PX') = \angle(SX_2, SX') = \varphi$ since S, X_2, P, X' are concyclic. □

Our first application of spiral similarity will be the proof of the so-called Simson[8] line.

Proposition 1.44 (Simson line). *Let ABC be a triangle and X a point in its plane. Denote by P, Q, R the projections of X to the sides BC, CA, AB, respectively. Then the points P, Q, R lie on a single line if and only if X lies on the circumcircle ω of the triangle ABC.*

Proof. First assume that $X \in \omega$. If X coincides with one of the vertices, we get the conclusion immediately. Also, if X is antipodal to one of the vertices (say A), then $Q = C$, $R = B$ and we are done. Otherwise, we look at right triangles XPC and XRA. The concyclicity of $ABCX$ gives

$$\angle(XA, AB) = \angle(XC, CB),$$

which means the triangles are directly similar. Now we consider spiral similarity centered at X which sends P to C and thus also R to A and denote by Q' the image of Q. Then as we preserve shape, $Q' \in AC$ and the collinearity of P, Q, and R follows from collinearity of their images C, Q', and A.

The argument may be reversed to show the "only if" part of the statement. □

[8]Robert Simson (1687–1768) was a Scottish mathematician and professor of mathematics at the University of Glasgow.

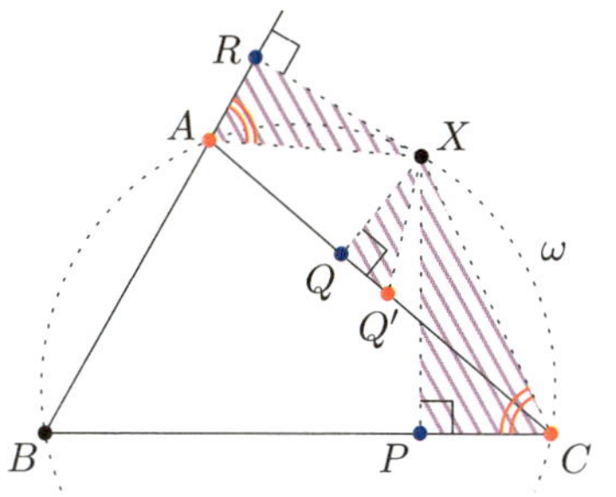

One thing to remember about spiral similarities is that they **come in pairs**. Whenever we come across a spiral similarity, there is always another one nearby.

Proposition 1.45. *Let $\mathcal{S}(S, k, \varphi)$ be a spiral similarity that maps A to A' and B to B'. Then:*

(a) $\triangle SAB \sim \triangle SA'B'$
(b) $\triangle SAA' \sim \triangle SBB'$.
(c) Spiral similarity $\mathcal{S}'(S, k', \varphi')$ maps A to B and A' to B' for suitable choice of k' and φ'.

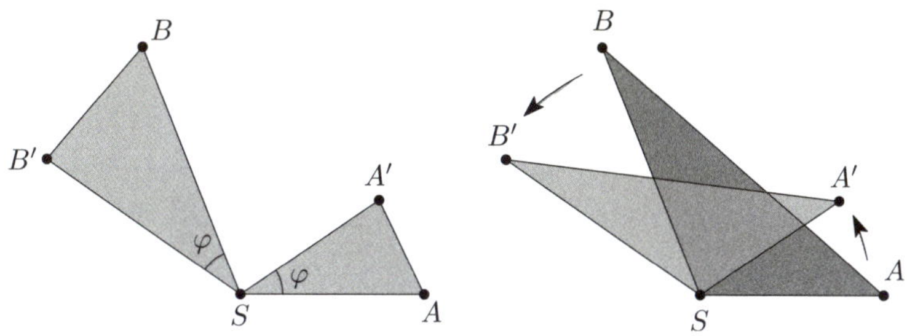

Proof. (a) Is immediate as $\mathcal{S}$ takes triangle SAB to triangle $SA'B'$.
(b) Follows from the definition of spiral similarity.
(c) Is a consequence of (b). □

Note that although these two spiral similarities share a center, they are not equal. They differ in dilation factor as well as in the angle of rotation.

This property of spiral similarity enables us to prove the famous theorem of Ptolemy[9] which provides a metric characterization of cyclic quadrilaterals.

Theorem 1.46 (Ptolemy's Inequality). *Let $ABCD$ be a quadrilateral. Denote the lengths of AB, BC, CD, DA by a, b, c, d, respectively, and its diagonals*

[9]Claudius Ptolemy (90–108 A.D.) was an Egyptian mathematician and astronomer.

AC, BD by e, f, respectively. Then

$$ac + bd \geq ef$$

and the equality holds if and only if $ABCD$ is cyclic.

Proof. Consider spiral similarity $\mathcal{S}$ with center C that sends D to B, and denote by A' the image of A under $\mathcal{S}$.

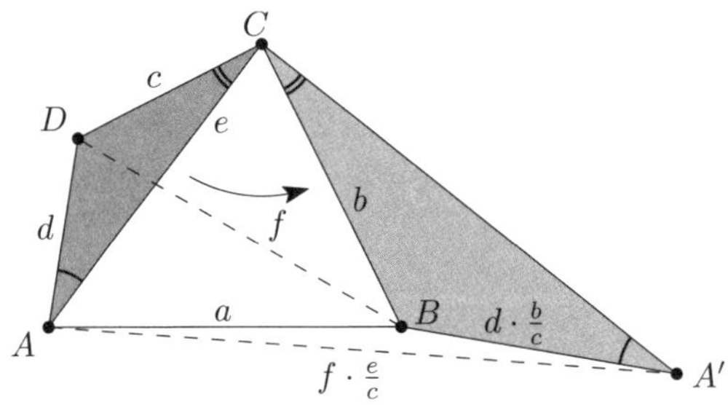

Since $\triangle CDA \sim \triangle CBA'$ (with factor $\frac{b}{c}$), we have $BA' = d \cdot \frac{b}{c}$. As spiral similarities come in pairs, we also have $\triangle CDB \sim \triangle CAA'$ (with factor $\frac{e}{c}$) and thus $AA' = f \cdot \frac{e}{c}$. From the triangle inequality applied to triangle ABA' we deduce

$$a + d \cdot \frac{b}{c} \geq f \cdot \frac{e}{c},$$

from which the result follows immediately. The equality occurs if and only if points A, B, A' are collinear, i.e. if $\angle CBA = 180^\circ - \angle A'BC = 180^\circ - \angle ADC$, which is equivalent to $ABCD$ being cyclic. □

Now we shall investigate, whether there exists a spiral similarity which sends two given points to two given points. The answer is positive.

Proposition 1.47. *Let A, B, A', B' be points in plane such that no three of them are collinear. Assume that the lines AB and $A'B'$ intersect at P. Then there exists unique spiral similarity that sends A to A' and B to B'. The center of this spiral similarity is the second intersection of the circumcircles of triangles $AA'P$ and $BB'P$.*

Proof. For S to be the center of the desired spiral similarity $\mathcal{S}(S, k, \varphi)$ that maps AB to $A'B'$, we need $\angle(SA, SA') = \angle(SB, SB') = \angle(AB, A'B') = \varphi$ (see Proposition 1.43(b)), implying that S has to belong to both circles circumscribed to triangles $AA'P$ and $BB'P$ (recall Proposition 1.18).

It remains to prove that triangles SAA' and SBB' are directly similar. We have already ensured $\angle(SA, SA') = \angle(SB, SB')$, and after we use the two circles, we obtain

$$\angle(A'A, AS) = \angle(A'P, PS) = \angle(B'P, PS) = \angle(B'B, BS),$$

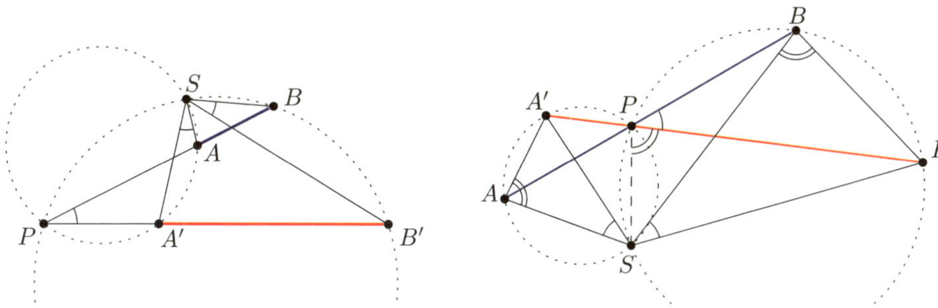

and we are done (AA).

If the circumcircles of triangles $AA'P$ and $BB'P$ happen to be mutually tangent, the spiral similarity degenerates to homothety with center P. □

The proposition does not apply to cases when some three of the four points are collinear. In these cases one of the circles becomes tangent to a corresponding line. Details are left to the reader.

The previous proposition can be restated so that it makes us more familiar with the configuration of two intersecting circles.

Proposition 1.48. *(a) Let SAB, $SA'B'$ be two directly similar triangles with circumcircles ω, ω', respectively. Then ω, ω' and the lines AA', BB' pass through a common point.*

(b) Let circles ω_1, ω_2 intersect at P and S. Then in the spiral similarity $\mathcal{S}$ with center S which takes ω to ω' point $A' \in \omega'$ is the image of $A \in \omega$ if and only if $P \in AA'$.

Proof. (a) If triangles SAB and $SA'B'$ have parallel sides, the common point is their center of homothety S. Suppose otherwise. Let $P = AA' \cap BB'$. Since S is the center of spiral similarity which sends A to B and A' to B', it is (by construction) the second intersection of the circumcircles of triangles ABP and $A'B'P$. Hence P lies on both ω and ω' and we may conclude.

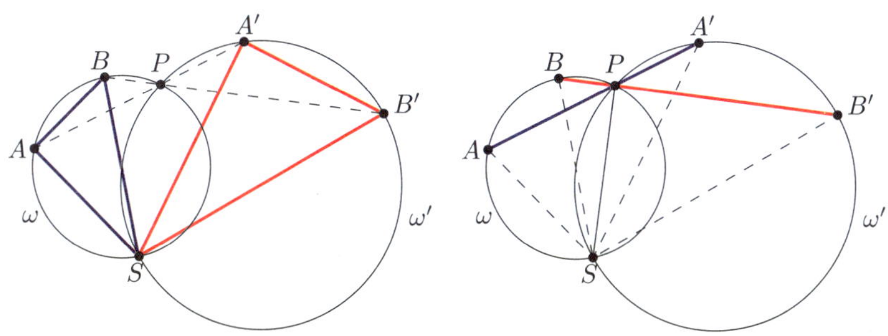

(b) First note that such spiral similarity exists. Now take points $A, B \in \omega$ and denote by $A', B' \in \omega'$ their images in $\mathcal{S}$. Then since $\triangle SAB \sim \triangle SA'B'$, (a) gives that AA' passes through P. We have proved that the (unique) image of A in $\mathcal{S}$ is the second intersection of AP and ω', so we are done. □

We have learned that every time we see two intersecting circles with some lines passing through one of the intersections, there is a spiral similarity to consider. And conversely, lines joining corresponding points in spiral similarity often pass through an intersection of two circles.

Example 1.17 (IMO 2006 shortlist). *Consider a convex pentagon $ABCDE$ such that*

$$\angle BAC = \angle CAD = \angle DAE, \qquad \angle CBA = \angle DCA = \angle EDA.$$

Let P be the point of intersection of the lines BD and CE. Prove that the line AP passes through the midpoint of the side CD.

Proof. Denote by ω_1, ω_2 the circumcircles of triangles BAC, DAE, respectively. Note that triangles BAC, CAD and DAE are mutually similar (AA).

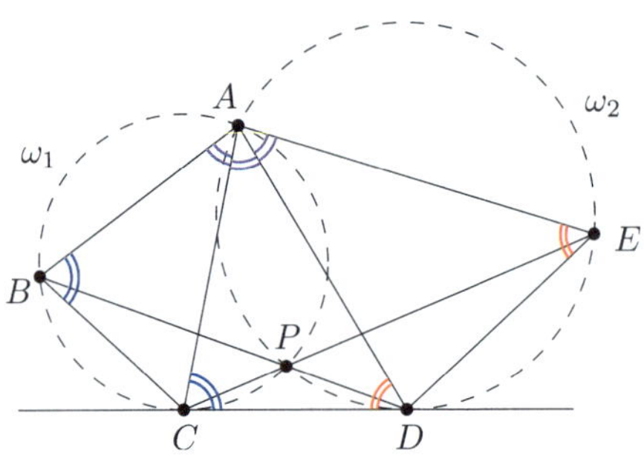

Consider the spiral similarity that maps triangle ABC to triangle ADE. Proposition 1.48(a) implies that P is also the second intersection of ω_1 and ω_2. From $\angle CBA = \angle DCA$ and $\angle ADC = \angle AED$ it follows that CD is tangent to both ω_1 and ω_2. Hence the midpoint of CD has equal power with respect to ω_1 and ω_2 (namely $\left(\frac{1}{2}CD\right)^2$) so it lies on their radical axis AP (consult Proposition 1.21 if needed). □

The following proposition can be proved by somewhat technical angle-chasing but equipped with the two previous propositions, we give an instant proof!

Proposition 1.49 (Miquel point of a quadrilateral). *Let $ABCD$ be a quadrilateral. Assume that rays BC and AD intersect at Q, and rays BA and CD*

intersect at R. Let ω_1, ω_2, ω_3, ω_4 be the circumcircles of triangles RAD, RBC, ABQ, CDQ, respectively. Then ω_1, ω_2, ω_3, ω_4 pass through a common point M. This point is called the Miquel point of the quadrilateral $ABCD$.

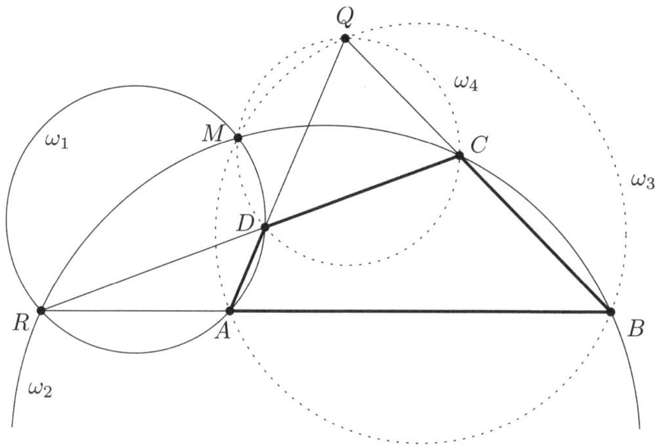

Proof. By Proposition 1.47, the second intersection M ($M \neq R$) of ω_1 and ω_2 is the center of the spiral similarity that maps A to D and B to C. By Proposition 1.45 it is also the center of the spiral similarity that maps A to B and D to C, so again by Proposition 1.47 it lies on ω_3 and ω_4. □

Example 1.18 (USAMO 2006). *Let $ABCD$ be a quadrilateral with nonparallel opposite sides and let E and F be points on the sides AD and BC, respectively, such that $AE/ED = BF/FC$. The ray FE meets the rays BA and CD at S and T, respectively. Prove that the circumcircles of triangles SAE, SBF, TCF, and TDE pass through a common point.*

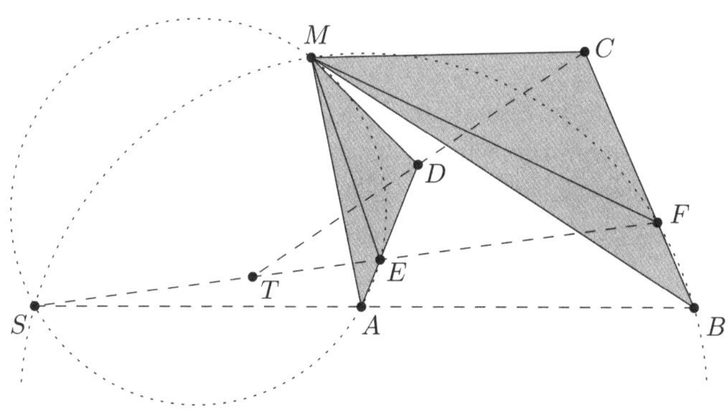

Proof. Let M be the center of the spiral similarity $\mathcal{S}$ that maps A to B and D to C. Then it takes AD to BC and as points E and F divide these segments in the same ratio, it also takes E to F.

Hence $\mathcal{S}$ maps segments AE to BF and ED to FC implying that M is the common Miquel point of quadrilaterals $ABFE$ and $EFCD$. Thus it lies on all the desired circles. □

For ample understanding of spiral similarity the next example is fundamental.

Example 1.19. *Two squares $ABCD$ and $A'B'C'D'$ (both labelled in counterclockwise order) are given in plane. Denote the midpoints of segments AA', BB', CC', DD' by A_1, B_1, C_1, D_1, respectively. Prove that $A_1B_1C_1D_1$ is a square.*

Proof. Consider spiral similarity $\mathcal{S}$ that maps A to A' and B to B'. The image of $ABCD$ under $\mathcal{S}$ is also a square. As it shares vertices A', B' with $A'B'C'D'$ and has the vertices labelled in the same order, it is in fact identical to $A'B'C'D'$. Hence $\mathcal{S}$ maps $ABCD$ to $A'B'C'D'$.

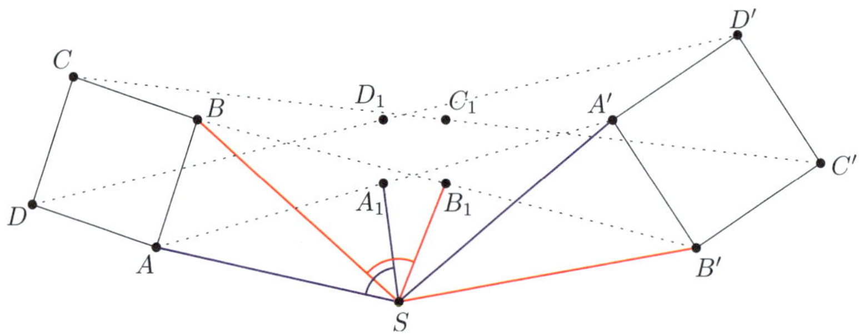

Now observe that by Proposition 1.45(b) triangles ASA', BSB', CSC', and DSD' are mutually similar. Since segments SA_1, SB_1, SC_1, SD_1 are medians in similar triangles, we have $\triangle ASA_1 \sim \triangle BSB_1 \sim \triangle CSC_1 \sim \triangle DSD_1$. Thus spiral similarity $\mathcal{S}'(S, \frac{SA_1}{SA}, \angle(SA, SA_1))$ maps $ABCD$ to $A_1B_1C_1D_1$ implying that $A_1B_1C_1D_1$ is indeed a square. □

Apparently, this example illustrates a more general concept. For example, we could replace two squares by any two directly similar figures. Also, we could divide the segments AA', BB', CC', DD' in any given ratio and the proposition would still hold. Loosely speaking, any "weighted average" of two directly similar (i.e. not necessarily equally oriented but labelled in the same direction) figures is a similar figure. To generalize yet further, we may even "average" more figures than two. The centroids, for instance, of the triangles formed by corresponding vertices of three mutually similar n-gons form again a similar n-gon. From now on we will refer to this principle as *Averaging Principle.*

Taking a bit different point of view we also see that if we join corresponding points of two directly similar figures and "glide" uniformly along these lines, then the shape of the figure is preserved. We choose to call this the *Gliding Principle.*

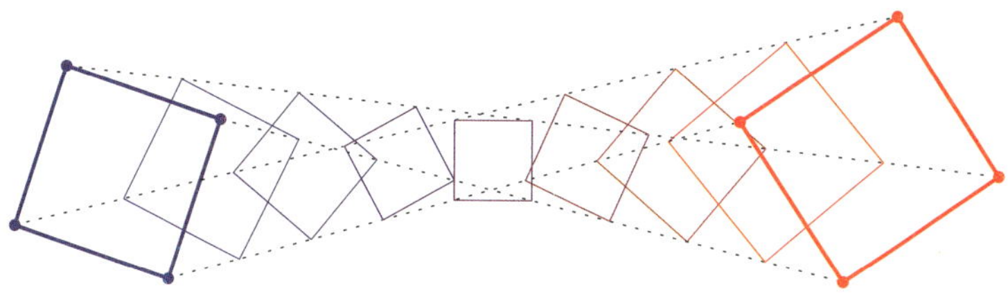

These principles generate tons of olympiad problems. Instead of a square, we can take a triangle with its orthocenter, segment with its midpoint, and so on. Every time we obtain a challenging problem!

The last example in this section only combines the ideas already discussed and fully exposes the power of spiral similarity.

Example 1.20. *Let circles ω_1, ω_2 centered at O_1, O_2, respectively, intersect at P and S. Points A, D on ω_1 and B, C on ω_2 are chosen such that segments AC and BD intersect at P. Denote the midpoints of AC, BD, O_1O_2 by M, N, O, respectively. Prove that O is the circumcenter of triangle MNP.*

Proof. Again by Proposition 1.48(b), S is the center of spiral similarity that sends A to C, D to B, ω_1 to ω_2 and thus also O_1 to O_2.

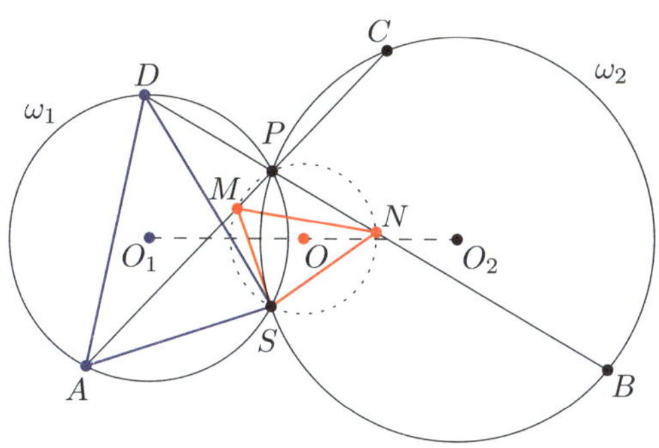

As triangle SAD glides to triangle SCB, its circumcenter O_1 glides along O_1O_2 and since P and S are symmetric about O_1O_2, its circumcircle at all times passes through P. Focusing on the situation in the middle of its way we realize that S, M, N, and P lie on a circle with center O. $\square$

Inversion

The most exotic geometric transformation we shall cover in this book is inversion. Unlike the transformations we have seen so far when applying inversion, figures may substantially change their shape. Yet, as we will see, inversion is an ultimately powerful tool in solving geometric problems.

Properties of inversion can be stated more efficiently if we introduce a point at infinity. We shall denote it as ∞ and we establish that it lies on each line. This extended plane is called the inversive plane.

Now let's disclose the definition. Given a circle ω with center I and radius $r > 0$ we define the image X' of point X under inversion about ω as follows:

(a) If $X = I$, then $X' = \infty$.
(b) If $X = \infty$, then $X' = I$.
(c) Otherwise, X' is such point on ray IX that $IX \cdot IX' = r^2$.

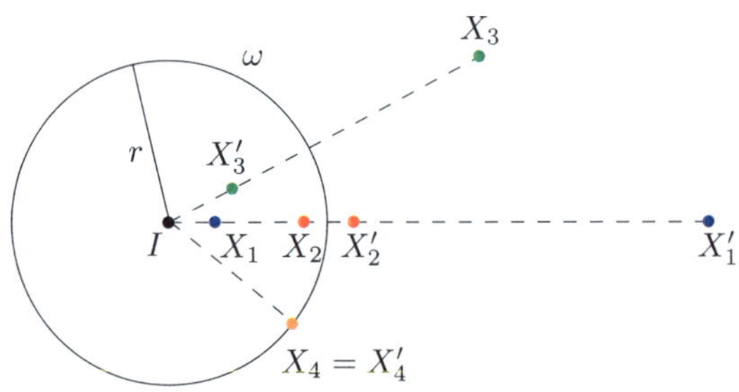

Observe that points inside ω (with $IX < r$) are mapped to the outside ($IX' > r$) and vice versa, while ω is left intact. Further, if we perform inversion about the same circle twice, we obtain identity mapping (nothing happens). In other words, X' is the image of X if and only if X is the image of X'.

Let's discover some further properties.

Proposition 1.50. *Let X be a point outside the circle ω centered at I. Let tangents from X touch ω at points A, B. Finally, denote by X' the midpoint of AB. Then X' is the image of X under inversion about ω.*

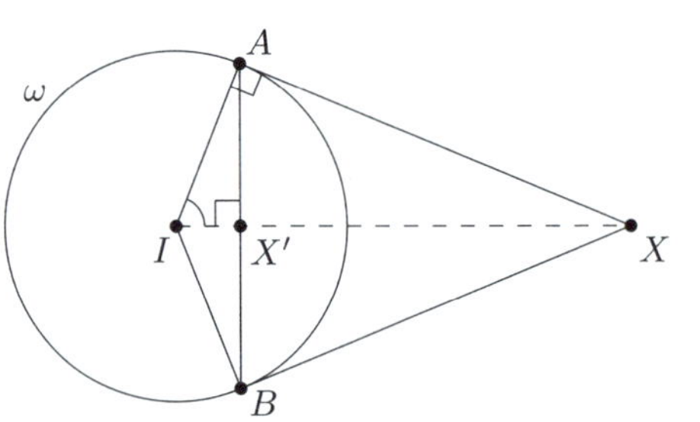

Proof. First note that by symmetry points I, X', X are collinear and $\angle IX'A = 90°$. Since AX is tangent to ω, we also have $\angle IAX = 90°$. Thus $\triangle IX'A \sim \triangle IAX$ (AA) and $IX' : IA = IA : IX$ which implies the desired result. $\square$

Soon, when we apply inversion to problems, the following property will be crucial. It will allow us to recalculate distances and angles in the inverted picture.

Proposition 1.51. *Let I, X, Y be pairwise distinct non-collinear points. Denote by X' and Y' the images of X and Y under inversion about a circle with center I and radius $r > 0$. Then $\triangle XIY \sim \triangle Y'IX'$ with ratio of similitude $\frac{X'Y'}{XY} = \frac{r^2}{IX \cdot IY}$. In particular,*

(a) $\angle XYI = \angle IX'Y'$.
(b) $X'Y' = XY \cdot \frac{r^2}{IX \cdot IY}$.
(c) $XY = X'Y' \cdot \frac{r^2}{IX' \cdot IY'}$.

Proof. By the definition of inversion we obtain $\frac{IX'}{IY} = \frac{r^2}{IY \cdot IX} = \frac{IY'}{IX}$, implying $\triangle XIY \sim \triangle Y'IX'$ (SAS). Parts (a) and (b) follow immediately, and for part (c), just recall that points X, Y are the images of X' and Y' and use (b). $\square$

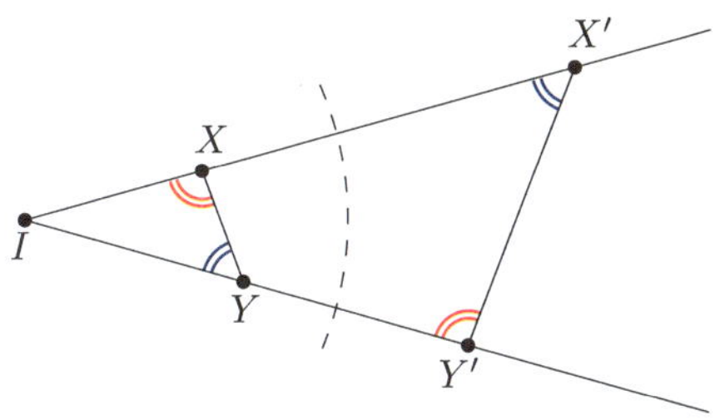

As we will see, the radius of inversion may often be chosen arbitrarily. In such case, we shall use the notion of inverting about a point. The radius will be considered to be equal to 1.

Now let's see what happens to lines and circles after inversion. The answer is surprisingly convenient!

Proposition 1.52. *Denote by ℓ' the image of line ℓ under inversion about I.*

(a) If $I \in \ell$, then $\ell' = \ell$.
(b) If $I \notin \ell$, then ℓ' is a circle with center O passing through I such that $OI \perp \ell$.

Proof. Part (a) is immediate since images of points from ℓ never leave this line and every point is attained (recall that I maps to ∞ and ∞ maps to I).

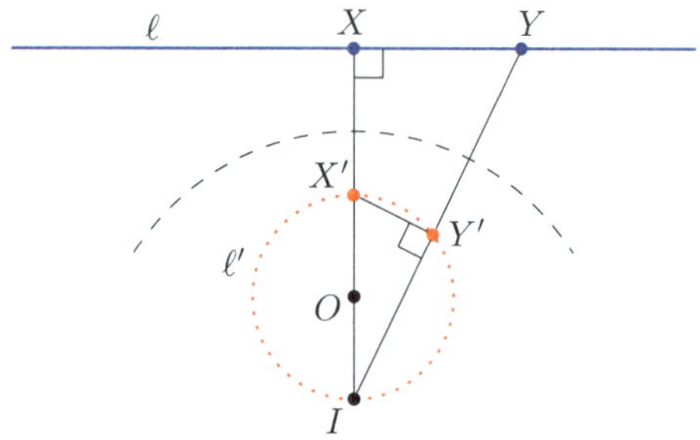

For part (b), denote by X the projection of I on ℓ, and let $Y \in \ell$, $Y \neq X$. Further, denote by X', Y' the images of X, Y under the inversion. As $\angle IY'X' = \angle IXY = 90^\circ$, point Y' lies on the circle with diameter IX'. It can be easily seen that each point of this circle is indeed attained (again recall that I maps to ∞ and ∞ maps to I). □

Proposition 1.53. *Denote by ω' the image of circle ω with center O under inversion about I.*

(a) If $I \in \omega$, then ω' is a line perpendicular to OI.
(b) If $I \notin \omega$, then ω' is a circle. Moreover, centers of ω and ω' are collinear with I.

Proof. Part (a) is essentially the same statement as Proposition 1.52(b).

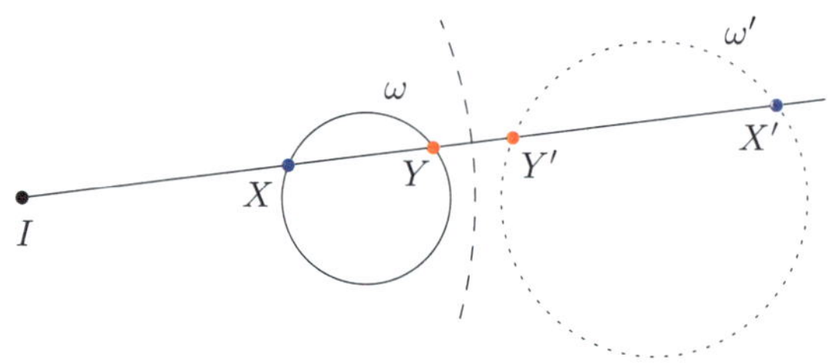

For part (b), let a line through I intersect ω at points X and Y and denote by X', Y' their respective images under inversion. Again we have

$$\frac{IX'}{IY} = \frac{1}{IY \cdot IX} = \frac{IY'}{IX},$$

thus if we consider homothety $\mathcal{H}(I, \frac{1}{IY \cdot IX})$, then points X', Y' are images of Y, X (in this order!). Since by Power of a Point the quantity $\frac{1}{IY \cdot IX}$ is constant as points X and Y vary on ω, the set ω' is just the image of ω in homothety $\mathcal{H}$ and inevitably it is a circle. Also, centers of ω and ω' are collinear with I.

□

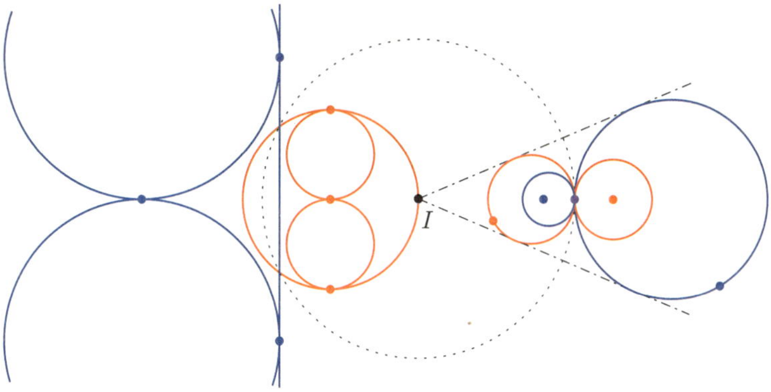

Which objects correspond under inversion about I?

It should be stressed that while circles are often mapped to circles, it is not true that their centers would be mapped to one another!

Mystery remains about how we apply inversion in problems. The idea is that we invert both the figure and the desired conclusion to obtain an equivalent problem. Very often (but not always!) this equivalent problem is far easier to solve.

As we will see in the first example, inverting about a point with many circles passing through it usually leads to a much simpler figure.

Example 1.21 (IMO 2003 shortlist). *Let Γ_1, Γ_2, Γ_3, Γ_4 be distinct circles such that Γ_1, Γ_3 are externally tangent at P, and Γ_2, Γ_4 are externally tangent at the same point P. Suppose that Γ_1 and Γ_2, Γ_2 and Γ_3, Γ_3 and Γ_4, Γ_4 and Γ_1 meet at A, B, C, D, respectively, and that all these points are different from P. Prove that*

$$\frac{AB \cdot BC}{AD \cdot DC} = \frac{PB^2}{PD^2}.$$

Proof. Invert about P (using standard notation for images $X \to X'$). Since

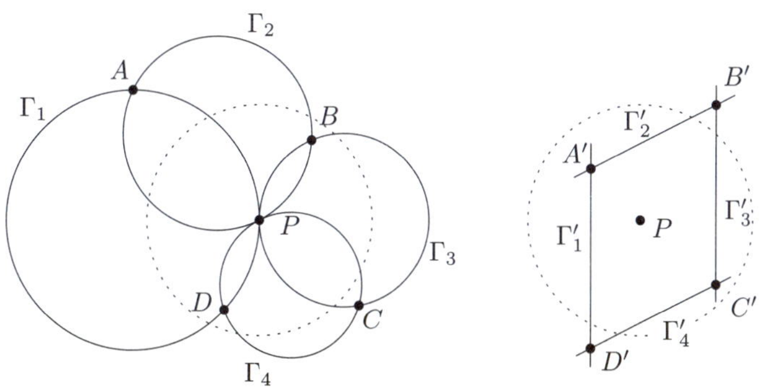

the circles Γ_1, Γ_3 are tangent at P, their centers are collinear with P and thus the circles will be transformed into a pair of parallel lines. The same argument applies for the circles Γ_2, Γ_4. Now observe that points A', B', C', D' are the intersection points of two pairs of parallel lines, and so they form (in this order) a parallelogram. In particular, we have $A'B' = C'D'$ and $B'C' = A'D'$. In terms of distances from the original picture this means (see Proposition 1.51(b))

$$\frac{AB}{PA \cdot PB} = \frac{CD}{PC \cdot PD}, \quad \frac{BC}{PB \cdot PC} = \frac{AD}{PD \cdot PA}.$$

Multiplying these two relations gives the result. □

The previous proof, although it is very short, does not give any guidelines as to how we should be proving metric identities after inversion. In the next example we will try to make it more understandable. The idea is that we perform some calculation (Proposition 1.51(c)) to see how the desired metric condition transforms into the inverted picture.

This time the strange constraints imposed on angles motivate the inversion. We hope they turn into something more approachable.

Example 1.22 (IMO 1996). *Let P be a point inside a triangle ABC such that*

$$\angle APB - \angle ACB = \angle APC - \angle ABC.$$

Let D, E be the incenters of triangles APB, APC, respectively. Show that the lines AP, BD, CE meet at a point.

Proof. We want to prove that the angle bisectors of $\angle PBA$ and $\angle ACP$ both intersect AP at the same point Z. By Angle Bisector Theorem applied to triangles PBA and PCA, this happens if and only if

$$\frac{AB}{PB} = \frac{AZ}{ZP} = \frac{AC}{PC}.$$

Hence it suffices to prove $\frac{AB}{PB} = \frac{AC}{PC}$ or $AB \cdot PC = AC \cdot PB$.

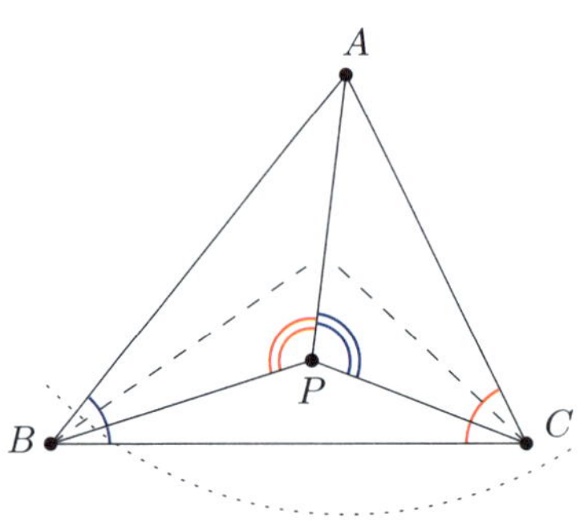

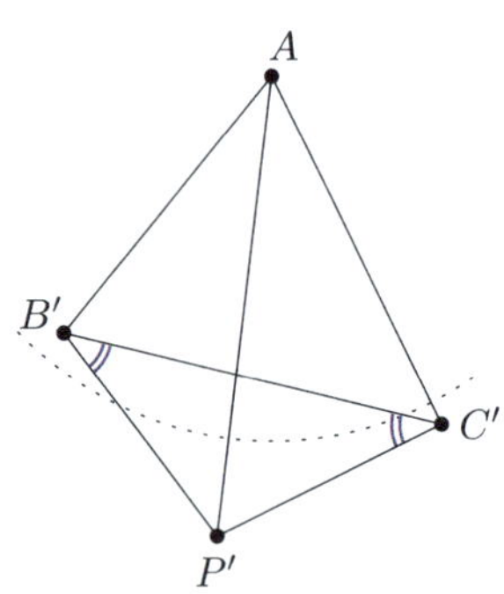

Invert about A. First, let's find out what happens to the metric relation we are proving. By Proposition 1.51(c) we are left to prove

$$\frac{1}{AB'} \cdot \frac{P'C'}{AP' \cdot AC'} = \frac{1}{AC'} \cdot \frac{P'B'}{AP' \cdot AB'}$$

or equivalently $P'C' = P'B'$. Now we transform the angular condition into the inverted picture. By Proposition 1.51(a) it is equivalent to

$$\angle P'B'A - \angle C'B'A = \angle P'C'A - \angle B'C'A,$$

or just $\angle P'B'C' = \angle P'C'B'$. So the triangle $P'C'B'$ is isosceles, which is exactly what we needed! □

To get a firm grasp of the previous technique, we strongly encourage the reader to try to solve the last example by inverting about P. The calculation is very similar.

Unlike the previous examples, this time we shall not use inversion to switch to a different problem. We consider some of its effects without leaving the given configuration. In such cases a good choice of inversion radius is often crucial.

Example 1.23 (Iran 1995). *Let M, N, P be the points where the incircle of scalene triangle ABC touches its sides BC, CA, AB, respectively. Prove that the orthocenter of triangle MNP, the incenter I of the triangle ABC and the circumcenter O of the triangle ABC are collinear.*

Proof. Note that I is the circumcenter of triangle MNP, so we are in fact proving that O lies on the Euler line (see Example 1.3) of triangle MNP. We invert about the incircle.

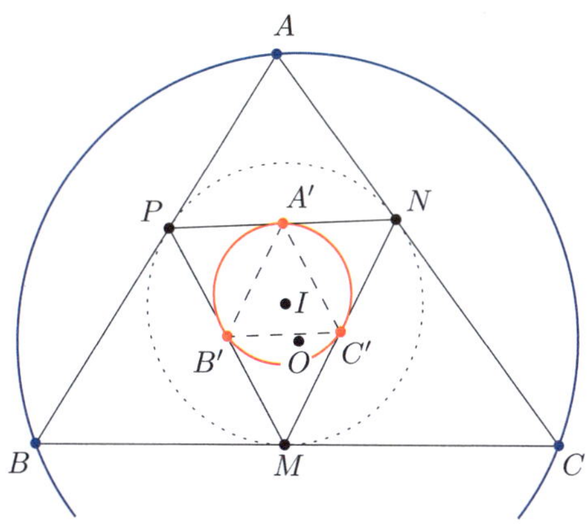

The images A', B', C' of points A, B, C are the midpoints of NP, MP and MN, respectively (see Proposition 1.50). Thus the circumcircle of triangle

ABC is taken to the circumcircle of triangle $A'B'C'$, i.e. the nine-point circle of triangle MNP. Denote the center of this nine-point circle by X.

As the center of a circle, the center of its image and the center of inversion are collinear, points O, X and I lie on a single line (but X is not the image of O, beware!). However, both I and X lie on the Euler line of triangle MNP (see Proposition 1.37), hence O lies there too. □

$\sqrt{bc}$-inversion

The last technique disclosed in this book connects inversion with antiparallel lines and triangle geometry. Given a triangle ABC we consider the transformation which first reflects point X over the A-angle bisector into X' and then inverts X' about A with radius $\sqrt{bc}$ into X''. We call X'' the image of X in $\sqrt{bc}$-inversion.

The seemingly complicated definition has many immediate and very pleasant consequences.

Proposition 1.54 ($\sqrt{bc}$-inversion properties)**.** *If we consider $\sqrt{bc}$-inversion in triangle ABC with A-angle bisector ℓ and circumcircle ω then the following holds:*

(a) B maps to C, C maps to B.
(b) ω maps to BC, BC maps to ω.
(c) Lines AX and AX' are isogonal for $X \neq A$.

Proof. As AB and AC are symmetric with respect to ℓ the image of B' lies on AC. Moreover, by the definition of inversion

$$AB \cdot AB' = AB \cdot AC,$$

thus indeed $AB' = AC$ and $B' = C$. For the same reason also C maps to B, which concludes the proof of (a).

For (b) just observe that the image of ω is a line passing through $B' = C$ and $C' = B$. Part (c) goes without saying.

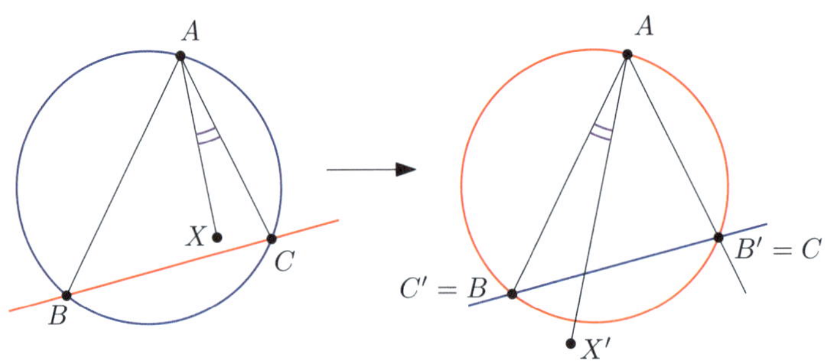

□

The power of $\sqrt{bc}$-inversion will be demonstrated on two examples.

Example 1.24. *Let ω be the circumcircle of triangle ABC. Circle ω_1 is inscribed in angle BAC and touches ω internally at T. Let D be the point of tangency of BC and the A-excircle. Show that $\angle BAT = \angle DAC$.*

Proof. We apply $\sqrt{bc}$-inversion and observe that ω_1' is still inscribed in $\angle BAC$ and as ω_1 touched ω internally, ω_1' touches BC externally, hence ω' is the A-excircle of triangle ABC. Thus T and D correspond in the $\sqrt{bc}$-inversion and the conclusion follows.

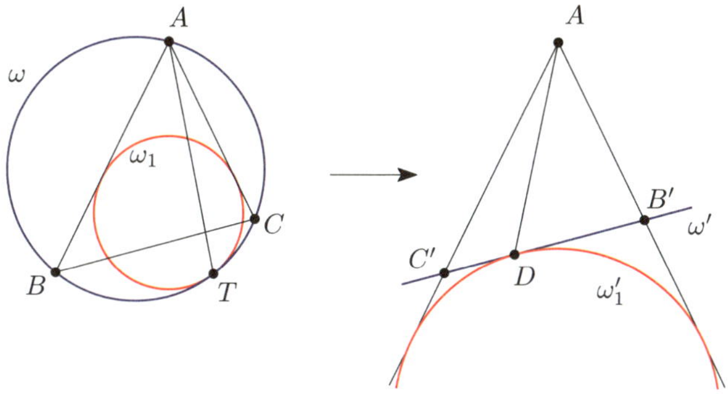

□

Example 1.25 (Serbia 2008). *Triangle ABC is given. Points D, E lie on the line AB such that $AD = AC$, $BE = BC$, and the points D, A, B, E are collinear in this order. Bisectors of internal angles at A and B intersect BC, AC at P and Q, respectively, and the circumcircle of triangle ABC at M and N, respectively. Line through A and the center O_1 of the circumcircle of triangle BME and line through B and the center O_2 of the circumcircle of triangle AND intersect at X. Prove that $CX \perp PQ$.*

Proof. We approach the point E metrically and use the Angle Bisector Theorem (see Proposition 1.10) to obtain

$$AE \cdot AQ = (a + c) \cdot \frac{bc}{a + c} = bc.$$

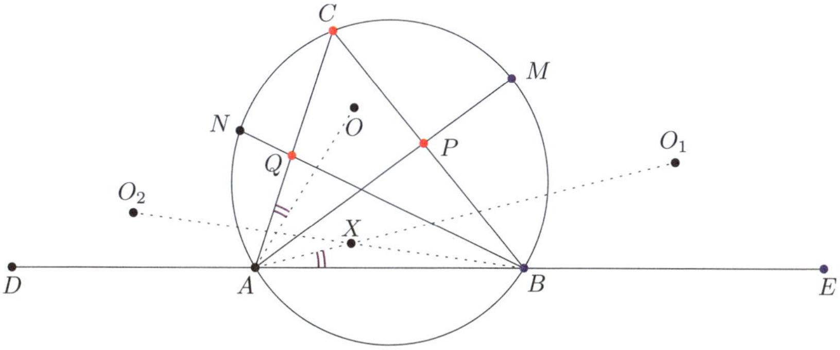

Then in $\sqrt{bc}$-inversion points E and Q correspond as well as points P and M. Thus the circumcircle of triangle BME corresponds to the circumcircle of triangle CPQ centered at O. Therefore, the line AO is isogonal to the line AO_1 in $\angle BAC$ (see Propositions 1.53(b) and 1.54(c)). Similarly, BO is isogonal to BO_2 in $\angle ABC$ and thus O and X are isogonal conjugates with respect to triangle ABC (see Proposition 1.26). Finally, in triangle CQP the line CX is isogonal with CO, thus it is the altitude (recall Proposition 1.17) and we are done. □

Chapter 2

Introductory Problems

1. Determine on which side is the driver's seat in the car depicted in the figure.

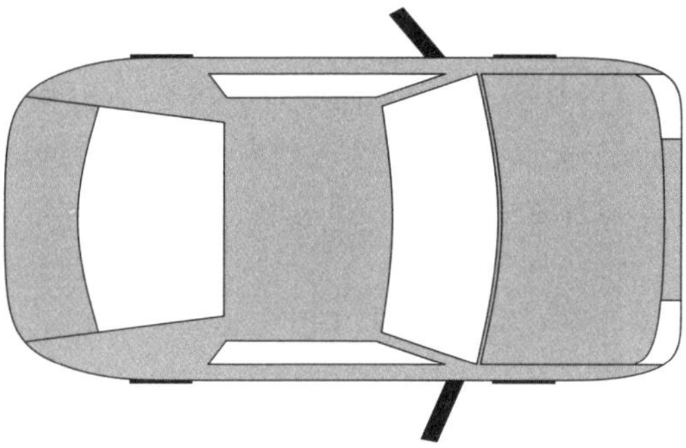

2. In right triangle ABC with hypotenuse BC let D be the foot of altitude from A. Show that

$$BD \cdot DC = DA^2, \quad BD \cdot BC = BA^2, \quad \text{and} \quad CD \cdot CB = CA^2.$$

3. Parallelogram $ABCD$ is given. The bisectors of $\angle A$ and $\angle B$ meet at E on the side CD. Prove that triangle AEB is right and that $AB = 2AD$.

4. Let AB be a fixed segment of length 6. Find the locus of the centers O of parallelograms $ABCD$ with $BC = 4$.

5. Through a fixed point O which is midway between two parallel lines we draw a variable line which intersects the parallel lines at points X, Y, respectively. Find the locus of points Z such that the triangle XYZ is equilateral.

6. Convex quadrilateral $ABCD$ is cut by lines connecting midpoints of its opposite sides into four pieces. Show the pieces may be rearranged to form a parallelogram.

7. Points D, E vary on the side BC of a triangle ABC such that $BD = CE$. Denote by M the midpoint of AD. Prove that all lines ME pass through a fixed point.

8. Show that the composition of two point reflections (i.e. performing one after the other) with distinct centers O_1 and O_2 results in a translation.

9. In acute triangle ABC let A_1, B_1, C_1 be the midpoints of the respective sides and A_0, B_0, C_0 the feet of respective altitudes. Prove that the length of the closed broken line $A_0B_1C_0A_1B_0C_1A_0$ equals the perimeter of triangle ABC.

10. Fixed circles ω_1, ω_2 of distinct radii are externally tangent at T. Consider all pairs of points $A \in \omega_1$, $B \in \omega_2$ such that $\angle ATB = 90°$. Show that all such lines AB pass through a fixed point.

11. Let ABC be a triangle. Denote by M, N, P the midpoints of its sides BC, CA, AB, respectively, and by J, K, L the incenters of the triangles APN, BMP, CNM, respectively.

    (a) Prove that $\triangle JKL \sim \triangle ABC$.

    (b) Prove that lines JM, KN, and LP are concurrent on the line IG, where I and G are the incenter and the centroid of triangle ABC, respectively.

12. Let ABC be a triangle with $AB < AC$. Denote by A_0 the foot of its A-altitude, by D the point of contact of the incircle with the side BC, by K the intersection of BC with the angle bisector of $\angle A$, and finally by M the midpoint of BC. Prove that points A_0, D, K, M are mutually different and lie on the line BC in this order.

13. Let ω be a fixed circle with center at O and radius R and let A be a fixed point outside the circle. Point X varies on ω so that A, O, and X are not collinear. Find the locus of the intersections Y of AX with the angle bisector of $\angle AOX$,

14. A variable point X runs along a semicircle ω with diameter AB ($X \neq A$, $X \neq B$). Let Y be such point on the ray XA that $XY = XB$. Find the locus of points Y.

15. A variable regular hexagon $ABCDEF$ has fixed point A and its center O is moving along a given line. Prove that the remaining five vertices also describe straight lines and that these lines are concurrent.

16. Let $ABCD$ be a cyclic quadrilateral and let H_d, H_c be the orthocenters of the triangles ABC and ABD, respectively.

    (a) Show that points A, B, H_d, H_c lie on a single circle.

    (b) Draw also H_a and H_b, the orthocenters of triangles BCD and CDA, and prove that $ABCD$ is congruent to $H_aH_bH_cH_d$.

17. Let D and E be the points of contact of the incircle of triangle ABC with its sides AB and AC, respectively. Also, let X be the circumcenter of triangle BIC, where I is the incenter of triangle ABC. Show that $\angle XDB = \angle XEC$.

18. Let ABC be a scalene acute-angled triangle with orthocenter H. Show that the Euler lines[1] of triangles BHC, CHA, AHB intersect at one point on the Euler line of triangle ABC.

19. Let ABC be a triangle and D the foot of its A-altitude. The line through A parallel to BC intersects the circumcircle ω of triangle ABC for the second time at E. Prove that line DE passes through the centroid of triangle ABC.

20. Let ω_1 and ω_2 be circles whose centers O_1, O_2 are 10 units apart and whose radii are 1 and 3 units. Find the locus of points M which are the midpoints of some segment XY, where $X \in \omega_1$ and $Y \in \omega_2$.

21. Let ω be a given circle. Points A, B, and C lie on ω such that ABC is an acute triangle. Points X, Y, and Z are also on ω such that $AX \perp BC$ at D, $BY \perp AC$ at E, and $CZ \perp AB$ at F. Show that the value of

$$\frac{AX}{AD} + \frac{BY}{BE} + \frac{CZ}{CF}$$

does not depend on the choice of A, B and C.

22. Let ABC be a triangle with $\angle A = 90°$ and let L be a point on BC. The circumcircles of the triangles ABL and ACL intersect AC and AB for the second time at M and N, respectively. Prove that $BM \perp CN$.

23. Triangle centers in other roles.

    Let ABC be an acute triangle. *Pedal triangle* of a point X is the triangle formed by the projections of X onto the triangle sides. Denote by I, O, H the incenter, circumcenter, and orthocenter of triangle ABC, respectively.

    (a) Prove that I is the circumcenter of its pedal triangle.

    (b) Prove that O is the orthocenter of its pedal triangle.

    (c) Prove that H is the incenter of its pedal triangle.

[1] For explanation see Example 1.3

24. Given a right triangle ABC, let $ABDE$ be a square erected outwards from its hypotenuse AB. Prove that the angle bisector of $\angle C$ bisects the area of the square $ABDE$.

25. Let $ABCD$ be a rhombus with a point P on the side BC and Q on the side CD such that $BP = CQ$. Prove that the centroid of the triangle APQ lies on the segment BD.

26. Let ABC be a triangle. Points M, N on its sides AB, AC, respectively, satisfy
$$\frac{BM}{AB} = 2 \cdot \frac{CN}{AC}.$$
The line perpendicular to MN passing through N intersects side BC at P. Prove that $\angle MPN = \angle NPC$.

27. Let ABC be a scalene triangle and denote by D the intersection of the external angle bisector at A with line BC. Prove that

    (a) $DB/DC = AB/AC$.

    (b) If we define points $E \in AC$ and $F \in AB$ also as feet of the respective external angle bisectors, then D, E, and F are collinear.

28. Let ABC be a scalene acute triangle. Draw points K, L, M, N such that $ABMN$ and $LBCK$ are congruent rectangles erected outwards from the triangle sides. Prove that lines AL, NK, MC are concurrent.

29. Let $ABCD$ be a convex quadrilateral whose diagonals intersect at right angle at O. Prove that the reflections of O across lines AB, BC, CD, DA are concyclic.

30. Let $ABCD$ be a cyclic quadrilateral and let I_1, I_2 be the incenters of the triangles ABC and ABD, respectively.

    (a) Show that the quadrilateral ABI_1I_2 is cyclic.

(b) Draw also I_3 and I_4, the incenters of triangles CDA and BCD, and prove that $I_1I_2I_3I_4$ is a rectangle.

31. Let M be the midpoint of the side BC of a triangle ABC. Point K on the segment AM satisfies $CK = AB$. Denote by L the intersection of CK and AB. Prove that triangle AKL is isosceles.

32. Let A_1, B_1, C_1 be the midpoints of the arcs BC, CA, AB of the circumcircle of triangle ABC (not containing A, B, C, respectively) and let A_2, B_2, C_2 be the tangency points of the incircle with BC, CA, AB, respectively. Prove that the lines A_1A_2, B_1B_2, C_1C_2 are concurrent.

33. Let ABC be a triangle with incenter I and A-excenter E. Further, let M be the midpoint of arc BC that does not contain A, and let $D = AI \cap BC$. Prove the following metric identities:

    (a) $AD \cdot AM = AB \cdot AC$.

    (b) $AI \cdot AE = AB \cdot AC$.

    (c) $MA \cdot ID = MI \cdot AI$.

34. Points M and N vary over the interiors of the sides AB and AC of a triangle ABC so that $BM/MA = AN/NC$. Prove that the circumcircles of the triangles AMN pass through another fixed point different from A.

35. A triangle ABC and a point D in its interior are given. Consider points E, F such that $\triangle AFB \sim \triangle CEA \sim \triangle CDB$, points B and E lie on different sides of the line AC, and points C and F lie on different sides of AB. Prove that $AEDF$ is a parallelogram.

36. Napoleon[2]'s Theorem

    Let ABC be a triangle and let BCD, CAE, ABF be equilateral triangles erected outwards from its sides. Show that the centroids A_1, B_1, C_1 of these equilateral triangles also form an equilateral triangle.

37. Let X be a point in the plane of triangle ABC such that

$$\frac{1}{XA} : \frac{1}{XB} : \frac{1}{XC} = a : b : c.$$

    Prove that the images of points A, B, C in inversion about X form an equilateral triangle.

38. Let $ABCD$ be a trapezoid such that $BC \parallel AD$ and $\angle CBA = 90^\circ$. Let M be a point on AB satisfying $\angle CMD = 90^\circ$. Let AK be an altitude in triangle DAM and BL an altitude in triangle MBC. Prove that the lines AK, BL, and CD are concurrent.

39. An angle with vertex V and a point A in its interior are given. Points X, Y lie on the respective rays of the angle such that $VX = VY$ and the sum $AX + AY$ is the minimal possible. Prove that $\angle XAV = \angle YAV$.

40. Let ABC be a triangle with $AB = AC$. Let K, L be the points on the sides AB, AC, respectively, such that $KL = BK + CL$. Let M be the midpoint of KL. The line through M parallel to AC intersects BC at N. Find the magnitude of the angle KNL.

41. Let ABC be a triangle and D the point of contact of the incircle ω with BC. Let DX be a diameter of ω. Show that if $\angle BXC = 90^\circ$, then $5a = 3(b + c)$.

---

[2]Napoleon Bonaparte (1769–1821) was a French amateur mathematician who sadly chose to win his fame in much less peaceful manner.

42. Given a triangle ABC with circumcenter O, orthocenter H, and circumradius R, prove that $OH < 3R$.

43. Circles ω_a, ω_b are internally tangent to a circle ω at distinct points A, B, respectively. Moreover, they are tangent to each other at T. Denote by P the second intersection of AT and ω. Show that BP is perpendicular to BT.

44. Let ABC be an acute-angled triangle with orthocenter H. Let A', B', C' be the images of A, B, C, respectively, under inversion about H. Prove that H is the incenter of triangle $A'B'C'$. What happens if triangle ABC is obtuse?

45. Circles ω_a, ω_b are internally tangent to a circle ω at distinct points A, B, respectively. Moreover, they are tangent to each other at T. Denote by P any intersection of ω and their common tangent through T. Let the lines PA, PB intersect ω_a, ω_b for the second time at X, Y, respectively. Show that XY is a common tangent of ω_a and ω_b.

46. Let ABC be a triangle and D the foot of the altitude from A. Let E and F lie on a line passing through D such that AE is perpendicular to BE, AF is perpendicular to CF, and E and F are different from D. Let M and N be the midpoints of the segments BC and EF, respectively. Prove that AN is perpendicular to NM.

47. Four distinct points P, Q, R, and S are given in plane, such that $PQRS$ is not a parallelogram. Find the locus of centers O of rectangles whose sidelines AB, BC, CD, and DA pass through P, Q, R, and S, respectively.

48. Let ω be a circle, BC its fixed chord, and A a variable point on its major arc BC. Let M be the point on the segment AB such that $AM = 2MB$ and let K be the projection of M onto AC. Show that point K moves along a circular arc.

49. In triangle ABC the line isogonal to the median is called the *symmedian.* Let ω be the circumcircle of triangle ABC.

    (a) If $\angle A \neq 90^\circ$ denote by T the intersection of tangents to ω at points B and C. Prove that line AT is the A-symmedian in triangle ABC.

    (b) Let the A-symmedian in triangle ABC meet ω for the second time at S. Prove that
    $$BS \cdot AC = CS \cdot AB.$$

    (c) If X is a point on the symmedian of the triangle ABC and we denote by X_b, X_c its projections on the sidelines AC and AB, then
    $$\frac{XX_b}{XX_c} = \frac{AC}{AB}.$$

50. Let A, B, C, and D be distinct points in the plane not lying on one circle. Each set of three points is inverted with respect to the fourth point. Show that the resulting four triangles are mutually similar.

51. Quadrilateral with escribed circle.

    Circle ω is inscribed in angle EAF and is tangent to AE at E and to AF at F. On the segments AE and AF choose points B and D, respectively. Let the tangents from B and D to ω (distinct from AE and AF) intersect at C. Show that:

    (a) $AB + BC = CD + DA$.

    (b) The incircles of triangles ABD and BCD touch BD at symmetric points with respect to the midpoint of BD.

52. Triangle ABC is inscribed in circle ω with radius R centered at O. Let I be the incenter of triangle ABC and r its inradius. Prove that $OI^2 = R^2 - 2Rr$.

53. Customizing inversion.

    (a) Let ω be a circle and I a point in its exterior. Prove that there exists a circle i centered at I such that ω is preserved in inversion about i.

(b) Let ω_1, ω_2, ω_3 be three circles with non-collinear centers, each outside of the other. Prove that there exists a circle i such that inversion about i preserves ω_1, ω_2, and ω_3.

Proof of (b). Let P be the radical center (see Proposition 1.22) of ω_1, ω_2, and ω_3. Since the circles lie outside each other, point P lies outside them too and $p(P, \omega_1) = p(P, \omega_2) = p(P, \omega_3) = p > 0$. As in the second proof of part (a) we conclude that the circle with center P and radius $\sqrt{p}$ has the desired property.

Chapter 3

Advanced Problems

1. In acute triangle ABC let E, F be the points of contact of the incircle with the sides AB, AC, respectively, and let L and M be the feet of B and C-altitudes. Show that the incenter I' of triangle ALM coincides with the orthocenter H' of triangle AEF.

2. In triangle ABC with $\angle BAC = 120°$, denote by D, E, F the intersections of the respective angle bisectors with the opposite sides BC, CA, AB. Find $\angle EDF$.

3. Let ABC be a triangle with $AB = AC$. Let D be the midpoint of BC, M the midpoint of AD and N the projection of D onto BM. Prove that $\angle ANC = 90°$.

4. Let ABC be an acute-angled triangle with $\angle A = 60°$ and $AB > AC$. Let I be its incenter.

   (a) If H is the orthocenter of triangle ABC, prove that

   $$2\angle AHI = 3\angle B.$$

   (b) If M the midpoint of AI, prove that M lies on the nine-point circle[1] of triangle ABC.

[1] For explanation see Theorem 1.37.

5. Quadrilateral $ABCD$ inscribed in a circle ω contains its center O in its interior. Let r and s be the lines obtained by reflecting AB with respect to the internal bisectors of $\angle CAD$ and $\angle CBD$, respectively. If P is the intersection of r and s, prove that OP is perpendicular to CD.

6. Let X be the foot of perpendicular from vertex B of the triangle ABC $(AB < AC)$ to the angle bisector of $\angle A$.

   (a) Let M, P be the midpoints of AB, BC, respectively. Prove that X lies on MP.

   (b) Let D, E be the points of contact of the incircle with sides BC, AC, respectively. Prove that X lies on the segment DE.

7. Let BK and CL be angle bisectors in an acute triangle ABC with incenter I (K lies on the side AC, L lies on the side AB). The perpendicular bisector of LC intersects the line BK at point M. Point N lies on the line CL such that NK is parallel to LM. Prove that $NK = NB$.

8. Circles ω_1, ω_2 with radii R_1 and R_2 are internally tangent at N (with ω_1 inside ω_2). Let K be an arbitrary point on ω_1. The tangent to ω_1 at K intersects ω_2 at A and B. Let M be the midpoint of the arc AB of ω_2 not containing point N. Prove that the circumradius R of triangle KBM does not depend on the choice of K.

9. The external common tangent of the circles Γ_1, Γ_2 with centers O_1, O_2 is tangent to them at distinct points A_1, A_2, respectively. The circle with diameter A_1A_2 meets Γ_1, Γ_2 for the second time at B_1, B_2, respectively. Prove that the lines A_1B_2, B_1A_2 and O_1O_2 are concurrent.

10. A circle passing through the vertex A of a parallelogram $ABCD$ intersects the segments AB, AC, AD for the second time at P, Q, R, respectively. Prove that

$$AP \cdot AB + AR \cdot AD = AQ \cdot AC.$$

11. A scalene triangle ABC with incenter I and $D = AI \cap BC$ satisfies $b + c = 2a$. Show that:

    (a) $GI \parallel BC$, where G is the centroid of triangle ABC.

    (b) $\angle OIA = 90^\circ$, where O is the circumcenter of triangle ABC.

    (c) Let E and F be the midpoints of AB and AC, respectively. Then I is the circumcenter of triangle DEF.

12. Points B, D, and C are collinear in this order and $BD < DC$. Find the locus of points X such that $\angle BXD = \angle DXC$.

13. Let ABC be a triangle and P a variable point on the arc AB of its circumcircle ω not containing point C. Let X, Y be the points on the rays BP, CP such that $BX = AB$ and $CY = AC$, respectively. Prove that all such lines XY pass through a fixed point independent of the choice of P.

14. Four circles ω, ω_a, ω_b, ω_c with the same radius are drawn in the interior of triangle ABC such that ω_a is tangent to the sides AB and AC, ω_b to BC and BA, ω_c to CA and CB, and ω is externally tangent to ω_a, ω_b, and ω_c. If the side lengths of triangle ABC are 13, 14, and 15, determine the radius of ω.

15. Broken circle.

    (a) Point P inside a parallelogram $ABCD$ satisfies $\angle BPC + \angle DPA = 180^\circ$. Prove that $\angle CBP = \angle PDC$.

    (b) Let $ABCD$ be a trapezoid with $AB \parallel CD$ and $AB > CD$. Points K and L lie on the line segments AB and CD, respectively, such that $\frac{AK}{KB} = \frac{DL}{LC}$. Suppose that there are points P and Q on the line segment KL satisfying $\angle APB = \angle DCB$ and $\angle CQD = \angle CBA$. Prove that the points P, Q, B, and C are concyclic.

16. Let ABC be an isosceles triangle with base BC. Let P be a point inside the triangle ABC such that $\angle CBP = \angle ACP$. Denote by M the midpoint of the base BC. Show that $\angle BPM + \angle CPA = 180^\circ$.

17. Let ABC be a non-right triangle with orthocenter H and circumcircle ω.

    (a) Let P be a point on ω. Prove that the reflections of P over the sides of the triangle ABC are collinear with H. Deduce that Simson line[2] of P with respect to triangle ABC bisects the segment PH.

    (b) Let ℓ be a line passing through H and denote by ℓ_a, ℓ_b, ℓ_c its reflections over the respective sides of the triangle ABC. Prove that ℓ_a, ℓ_b, ℓ_c pass through a common point on ω.

18. Circles ω_a, ω_b are externally tangent at T and their common external tangent ℓ is tangent to them at A, B, respectively. Let ω be a circle inscribed in the curvilinear triangle ABT and denote by O its center and by r its radius. Prove that $OT \leq 3r$.

19. Let ABC be a triangle inscribed in circle ω and denote by R, r, r_a, r_b, r_c its circumradius, inradius, and the respective exradii.

    (a) Denote by M the midpoint of the side BC and by N the midpoint of arc BC of ω containing vertex A. Prove that

    $$MN = \frac{1}{2}(r_b + r_c).$$

    (b) Prove that

    $$r_a + r_b + r_c = 4 \cdot R + r.$$

    (c) Let D, E, F be the midpoints of arcs BC, CA, AB of ω not containing vertices A, B, C, respectively. Prove that the perimeter of the hexagon $AFBDCE$ is at least $4(R + r)$.

[2]For explanation see Proposition 1.44.

20. Circles ω_1, ω_2, and ω_3 are given in the plane, every one outside the others. Circle ω is tangent to them externally at A_1, A_2, A_3, respectively, and circle Ω is tangent to them internally at B_1, B_2, B_3, respectively. Prove that lines A_1B_1, A_2B_2, and A_3B_3 are concurrent.

21. Points K, L on the side BC of a triangle ABC satisfy $\angle BAK = \angle CAL < \frac{1}{2}\angle A$. Let ω_1 be any circle tangent to the lines AB and AL, let ω_2 be any circle tangent to the lines AC and AK, and suppose that ω_1 and ω_2 intersect at P and Q. Prove that $\angle PAC = \angle QAB$.

22. An acute-angled triangle ABC is given. A circle passing through A and the triangle's circumcenter O intersects AB and AC at points P and Q, respectively. Prove that the orthocenter of the triangle POQ lies on the line BC.

23. Let O be the circumcenter of a triangle ABC. Points M and N are chosen on the sides AB and AC, respectively, so that $\angle NOM = \angle A$. Prove that the perimeter of triangle MAN is not less than the length of the side BC.

24. Let ABC be a scalene triangle with orthocenter H and incenter I. Line ℓ_a is perpendicular to the bisector of $\angle A$ and passes through the midpoint of BC. Lines ℓ_b and ℓ_c are defined analogously. Show that the circumcenter O_1 of triangle formed by these lines lies on the line IH.

25. Let ω_a, ω_b be two circles that are externally tangent at T and internally tangent to circle ω at A, B, respectively. Let S be one of the intersections of the common tangent of ω_a, ω_b at T with ω. Line AS intersects ω_a again at C and BS intersects ω_b again at D. Line AB intersects ω_a again at E and ω_b again at F. Prove that lines ST, CE, DF are concurrent.

26. Shortest paths.

    (a) Let ℓ be a line and A, B two points on the same side of it. For what point $L \in \ell$ is $AL + LB$ minimal?

(b) Let ABC be an acute-angled triangle. Among all the triangles DEF with vertices D, E, F on the sides BC, CA, AB, respectively, one has minimal perimeter. Find which one.

27. Circles ω_1, ω_2 inscribed in a given circular sector with endpoints A, B are externally tangent at T. Denote by ℓ their common internal tangent.

    (a) Prove that ℓ passes through a fixed point independent of the position of ω_1, ω_2.

    (b) Let C be the intersection of ℓ with arc AB. Prove that T is the incenter of triangle ABC.

28. Let $ABCD$ be a fixed convex quadrilateral with $BC = DA$ and BC not parallel to DA. Let two variable points E and F lie on the sides BC and DA, respectively, and satisfy $BE = DF$. The lines AC and BD meet at P, the lines BD and EF meet at Q, the lines EF and AC meet at R. Prove that the circumcircles of the triangles PQR, as E and F vary, have a common point other than P.

29. Let $ABCD$ be a quadrilateral inscribed in a semicircle ω with diameter AB and center O. Lines CD and AB intersect at M. Let K be the second point of intersection of the circumcircles of triangles AOD and BOC. Prove that $\angle MKO = 90^\circ$.

30. Let AB be a segment and C its midpoint. Circle ω_1 which passes through A and C intersects circle ω_2 which passes through B and C at two different points C and D. Point P is the midpoint of arc AD of circle ω_1 which does not contain C. Similarly, point Q is the midpoint of arc BD of circle ω_2 which does not contain C. Prove that $PQ \perp CD$.

31. Let BC be a fixed chord of the circle ω with radius R and let A vary on the major arc BC of ω forming an acute triangle ABC with $\angle A \neq 60^\circ$ and orthocenter H.

(a) Show that the mirror images H' of H over the A-angle bisector run along a circle.

(b) Show that the projections X of H on the A-angle bisector also run along a circle.

32. In acute triangle ABC inscribed in circle ω, let A' be the projection of A onto BC and B', C' the projections of A' onto AC, AB, respectively. Line $B'C'$ intersects ω at X and Y and line AA' intersects ω for the second time at D. Prove that A' is the incenter of triangle XYD.

33. Given a triangle ABC, let B_1, B_2, and C_1, C_2 be points on the sides AB and AC, respectively, such that $BB_1/BB_2 = CC_1/CC_2$. Prove that the orthocenters of triangles ABC, AB_1C_1, and AB_2C_2 are collinear.

34. Let ABC be a scalene triangle. The angle bisector of $\angle A$ intersects the side BC at D and the circumcircle Ω of triangle ABC at A and E. Circle ω with diameter DE cuts Ω again at F. Prove that AF is the symmedian[3] of triangle ABC.

35. Let ABC be a triangle, let K be the midpoint of the side AB and L the midpoint of the side AC. Let P be the second intersection of the circumcircles of triangles ABL and AKC. Let Q be the second intersection of AP and the circumcircle of triangle AKL. Prove that $2AP = 3AQ$.

36. An angle of fixed magnitude φ revolves about its fixed vertex A and meets a fixed line ℓ at points B and C. Prove that the circumcircles of triangles ABC are all tangent to a fixed circle.

[3]For explanation see Introductory Problem 49.

37. Let ABC be a triangle and denote its circumcircle centered at O by ω. Points M and N lie on the sides AB and AC, respectively. The circumcircle of triangle AMN intersects ω for the second time at Q. Let P be the intersection point of MN and BC. Prove that PQ is tangent to ω if and only if $OM = ON$.

38. Let $ABCD$ be a cyclic quadrilateral. The projections of the intersection of its diagonals P to the sides AB and CD are E, F, respectively. Show that the line EF is perpendicular to the line through the midpoints K and L of the sides of BC and DA, respectively.

39. Given a triangle ABC with incenter I and circumcircle Γ, let AI intersect Γ again at D. Let E be a point on the arc BDC, and F a point on the segment BC, such that $\angle BAF = \angle EAC < \frac{1}{2}\angle BAC$. If G is the midpoint of IF, prove that lines EI and DG intersect on Γ.

40. Let $ABCDE$ be a regular pentagon. Find the minimum possible value of

$$\frac{PA + PB}{PC + PD + PE}$$

where P is any point in the plane.

41. Let ABC be an A-isosceles triangle inscribed in circle Ω. Arbitrary circles ω_b, ω_c inscribed in the minor circular segments AC, AB of Ω are tangent to Ω at B', C', respectively. One of the common external tangents of ω_b and ω_c intersects the sides AC, AB at P, Q, respectively. Prove that lines $B'P$ and $C'Q$ intersect on the angle bisector of $\angle BAC$.

42. Let ABC be a triangle and let ω be its incircle. Denote by D_1 and E_1 the points where ω is tangent to the sides BC and AC, respectively. Denote by D_2 and E_2 the points on sides BC and AC, respectively, such that $CD_2 = BD_1$ and $CE_2 = AE_1$, and denote by P the point of intersection of segments AD_2 and BE_2. Circle ω intersects segment AD_2 at two points, the closer of which to the vertex A is denoted by Q. Prove that $AQ = D_2P$.

43. Let ABC be an acute, scalene triangle, and let M, N, and P be the midpoints of BC, CA, and AB, respectively. Let the perpendicular bisectors of AB and AC intersect ray AM in points D and E, respectively, and let lines BD and CE intersect in point F, inside triangle ABC. Prove that points A, N, F, and P all lie on one circle.

44. Let MN be a line parallel to the side BC of a triangle ABC, with M on the side AB and N on the side AC. The lines BN and CM meet at point P. The circumcircles of triangles BMP and CNP meet at two distinct points P and Q. Prove that $\angle BAQ = \angle CAP$.

45. Let $ABCDEF$ be a convex hexagon such that $\angle B + \angle D + \angle F = 360°$ and
$$\frac{AB}{BC} \cdot \frac{CD}{DE} \cdot \frac{EF}{FA} = 1.$$
Prove that
$$\frac{BC}{CA} \cdot \frac{AE}{EF} \cdot \frac{FD}{DB} = 1.$$

46. In acute scalene triangle ABC with orthocenter H, denote by α', β', and γ' the magnitudes of angles $180° - \angle A$, $180° - \angle B$, and $180° - \angle C$, respectively. Points H_a, H_b, and H_c in the interior of triangle ABC satisfy
$$\begin{aligned} \angle BH_aC &= \alpha', & \angle CH_aA &= \gamma', & \angle AH_aB &= \beta', \\ \angle CH_bA &= \beta', & \angle AH_bB &= \alpha', & \angle BH_bC &= \gamma', \\ \angle AH_cB &= \gamma', & \angle BH_cC &= \beta', & \angle CH_cA &= \alpha'. \end{aligned}$$
Prove that the points H, H_a, H_b, H_c are concyclic.

47. Let ABC be an acute-angled triangle with $AB \neq AC$. Let H be the orthocenter of triangle ABC, and let M be the midpoint of the side BC. Let D be a point on the side AB and E a point on the side AC such that $AE = AD$ and the points D, H, E lie on the same line. Prove that the line HM is perpendicular to the common chord of the circumscribed circles of the triangles ABC and ADE.

48. Let $ABCD$ be a cyclic quadrilateral. Draw all excenters of triangles ABC, BCD, CDA, and DAB. Show that these twelve points lie on the perimeter of a rectangle.

49. Let ABC be a triangle, H its orthocenter, O its circumcenter, and R its circumradius. Let D be the reflection of the point A across the line BC, let E be the reflection of the point B across the line CA, and let F be the reflection of the point C across the line AB. Prove that the points D, E and F are collinear if and only if $OH = 2R$.

50. Points A_1, B_1, C_1 are chosen on the sides BC, CA, AB of a triangle ABC, respectively. The circumcircles of triangles AB_1C_1, BC_1A_1, CA_1B_1 intersect the circumcircle ω of triangle ABC for the second time at points A_2, B_2, C_2, respectively. Points A_3, B_3, C_3 are symmetric to A_1, B_1, C_1 with respect to the midpoints of the sides BC, CA, AB, respectively. Prove that the triangles $A_2B_2C_2$ and $A_3B_3C_3$ are similar.

51. The incircle ω of the acute-angled triangle ABC is tangent to its side BC at a point K. Let AD be an altitude of triangle ABC, and let M be its midpoint. If N is the common point of the circle ω and the line KM (distinct from K), then prove that the incircle ω and the circumcircle ω' of triangle BCN are tangent to each other at the point N.

52. Let ABC be a triangle inscribed in the circle ω. Point D is chosen on the side BC. Circle ω_1 is tangent to the segment BD at K, to the segment AD at L and to ω at T. Prove that the line KL passes through the incenter I of the triangle ABC.

53. Let $ABCD$ be a convex quadrilateral with BA different from BC. Denote the incircles of triangles ABC and ADC by ω_1 and ω_2, respectively. Suppose that there exists a circle ω tangent to ray BA beyond A and to the ray BC beyond C, which is also tangent to the lines AD and CD. Prove that the common external tangents to ω_1 and ω_2 intersect on ω.

54. Let XVY be a fixed angle with magnitude φ and P a point inside of it. A variable angle APB with $A \in VX$ and $B \in VY$ and fixed magnitude $\psi > \varphi$ and $2\psi - \varphi \neq 180°$ revolves around P. Show that there exists point Q in the plane ($Q \neq P$, $Q \neq V$) such that the angle AQB is fixed.

Chapter 4

Solutions to Introductory Problems

1. [Sharygin Geometry Olympiad 2007] Determine on which side is the driver's seat in the car depicted in the figure.

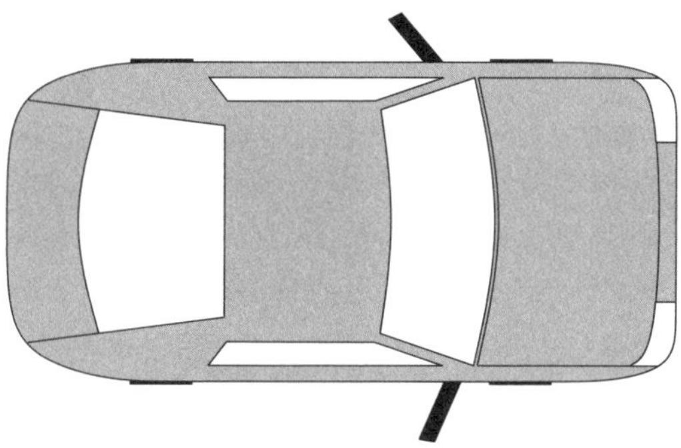

Proof. Taking the positions of the rear-view mirrors into account, the driver's seat is certainly on the right!

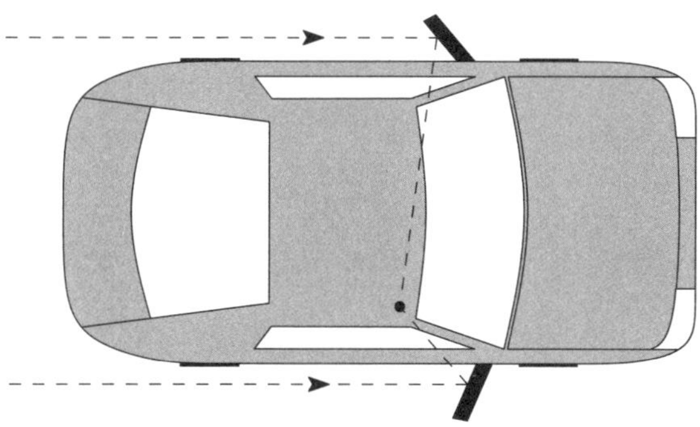

2. In right triangle ABC with hypotenuse BC let D be the foot of altitude from A. Show that

$$BD \cdot DC = DA^2, \quad BD \cdot BC = BA^2, \quad \text{and} \quad CD \cdot CB = CA^2.$$

First Proof. We claim that the three right triangles ABC, DBA, and DAC are pairwise similar. Indeed, since

$$\angle DBA = 90^\circ - \angle ACD = \angle DAC,$$

all the similarities follow (AA).

From $\triangle BDA \sim \triangle ADC$, we learn that $BD/DA = DA/DC$ which rewrites as $BD \cdot DC = DA^2$.

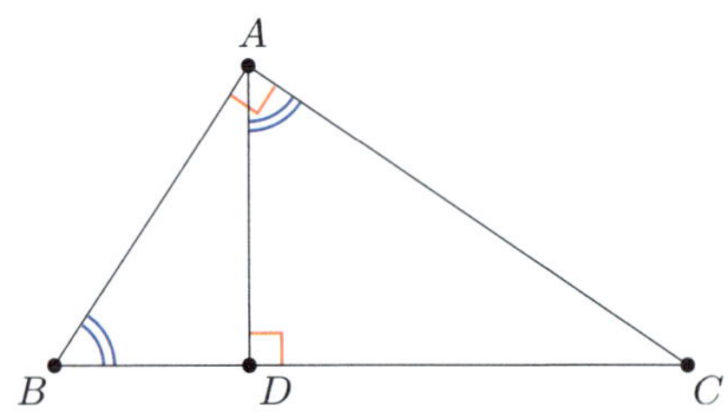

And $\triangle BDA \sim \triangle BAC$ yields $BD/BA = BA/BC$ which proves the second relation. The third one is proved analogously.

Second Proof. Since $\angle BAC$ is right, BC is a diameter of the circumcircle of triangle ABC. Hence the second point where AD meets this circumcircle is the reflection A' of A across BC and $DA = DA'$. Hence the first equality is just the power of D with respect to the circumcircle of triangle ABC.

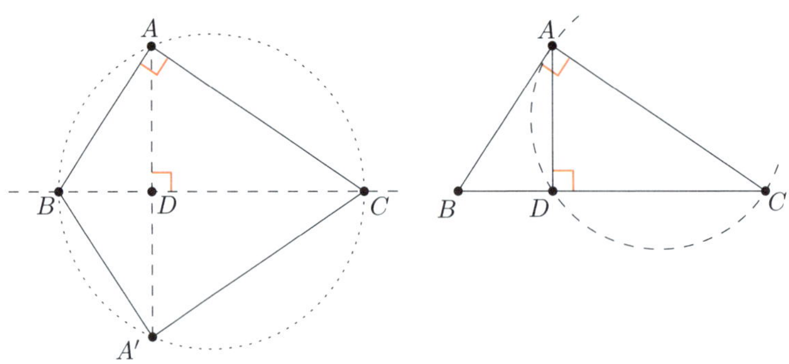

Since the line BA is perpendicular to the diameter of the circumcircle of triangle ACD, it is its tangent at A. Hence the second equality is just the power of B with respect to the circumcircle of triangle ACD.

Similarly, the third is the power of C with respect to the circumcircle of triangle ABD.

3. Parallelogram $ABCD$ is given. The bisectors of $\angle A$ and $\angle B$ meet at E on the side CD. Prove that triangle AEB is right and that $AB = 2AD$.

First Proof. First, since the lines AD and BC are parallel, the angle bisectors of the supplementary angles DAB and ABC are perpendicular. Indeed,

$$\angle EAB + \angle ABE = \frac{1}{2}(\angle DAB + \angle ABC) = \frac{1}{2} \cdot 180^\circ = 90^\circ$$

and $\angle BEA = 180^\circ - (\angle EAB + \angle ABE) = 90^\circ$.

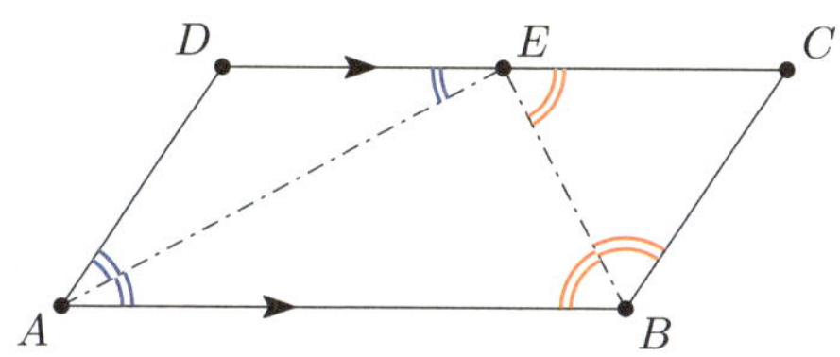

As for the second part, using the fact that lines AB and CD are parallel we learn

$$\angle DEA = \angle EAB = \frac{1}{2}\angle A = \angle DAE$$

implying that triangle DAE is D-isosceles and $DE = AD$. Likewise, we get $EC = BC$ and finally we may conclude by

$$AB = DC = DE + EC = AD + BC = 2AD.$$

Second Proof. Let line through E parallel to AD and BC intersect AB at M. Both $AMED$ and $MBCE$ are then parallelograms in which a diagonal coincides with the angle bisector so they are in fact rhombi.

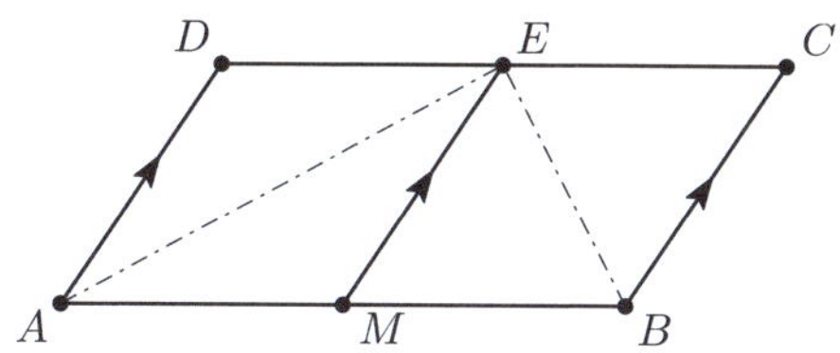

Since the rhombi share a side, they are congruent and $AB = 2AD$. Also, $ME = MA = MB$ implies that M is the circumcenter of triangle ABE and hence $\angle AEB = 90^\circ$.

4. Let AB be a fixed segment of length 6. Find the locus of the centers O of parallelograms $ABCD$ with $BC = 4$.

Solution. Since $BC = 4$ is fixed, the locus of vertices C of all such parallelograms is a circle ω with center B and radius 4 (without its two intersections with the line AB).

Now it suffices to realize that point O, being the center of the parallelogram $ABCD$, is the midpoint of the diagonal AC. Denoting the midpoint of AB by M and considering the homothety with center A and factor $\frac{1}{2}$ we therefore obtain that as C runs along ω, point O traces a circle with center M and radius $\frac{1}{2} \cdot 4 = 2$.

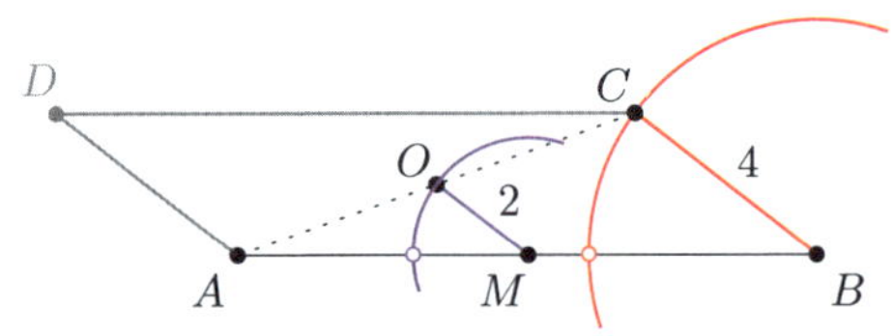

The sought-after locus is the circle with center M and radius 2 without its two intersections with the line AB.

5. Through a fixed point O which is midway between two parallel lines we draw a variable line which intersects the parallel lines at points X, Y, respectively. Find the locus of points Z such that the triangle XYZ is equilateral.

Solution. Since O lies midway between the two parallel lines, it is the midpoint of the segment XY and all the triangles XOZ have the same shape – namely a half of the equilateral triangle, i.e. the "30-60-90" triangle. Point Z is thus the image of X in spiral similarity $\mathcal{S}$ with fixed center O, factor $\sqrt{3}$, and angle $\pm 90°$.

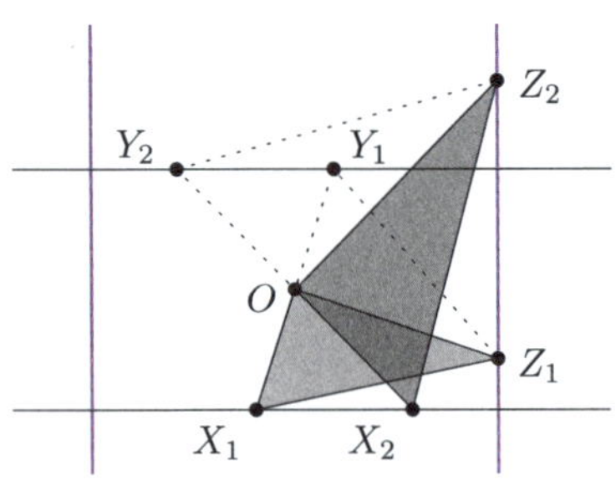

As X runs along one of the parallel lines, the locus of Z consists of its image(s) in $\mathcal{S}$, i.e. a pair of lines perpendicular to the given ones and with distance from point O multiplied by $\sqrt{3}$.

6. Convex quadrilateral $ABCD$ is cut by lines connecting midpoints of its opposite sides into four pieces. Show the pieces may be rearranged to form a parallelogram.

Proof. Denote the midpoints of the sides AB, BC, CD, DA by K, L, M, N, respectively, and the pieces by vertices A, B, C, D by $\mathcal{A}$, $\mathcal{B}$, $\mathcal{C}$, $\mathcal{D}$, respectively. We rearrange them into a parallelogram with sides parallel to KM and LN.

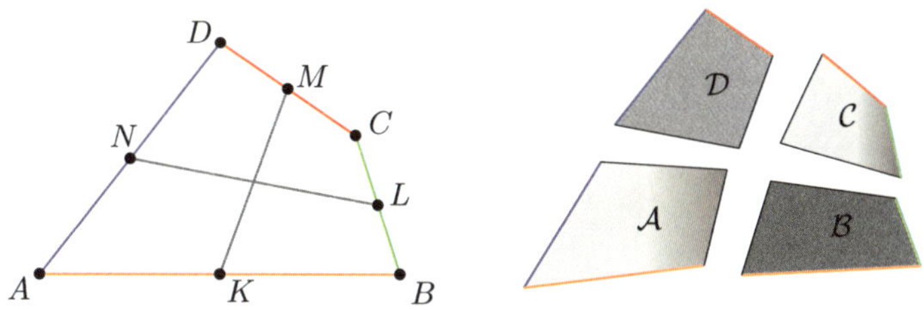

First we interchange the pieces $\mathcal{B}$ and $\mathcal{D}$, then we rotate each of the pieces $\mathcal{A}$ and $\mathcal{C}$ by $180°$, and finally we glue all the four pieces together by one common vertex.

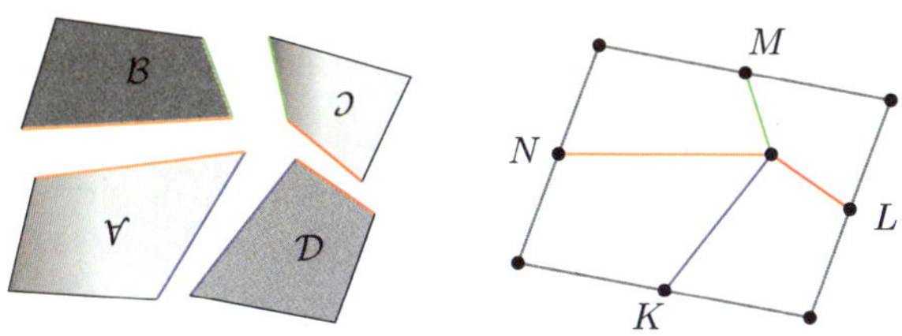

To make sure that such operation produces a parallelogram, observe that the angles in the middle add up to $\angle A + \angle B + \angle C + \angle D = 360°$, at all places we glue together equal segments (K, L, M, N were the midpoints) and finally as every piece was either translated or rotated by $180°$, the directions of all their sides were preserved. The resulting figure is thus a quadrilateral with pairs of opposite sides parallel to KM and LN, respectively, i.e. a parallelogram.

7. Points D, E vary on the side BC of a triangle ABC such that $BD = CE$. Denote by M the midpoint of AD. Prove that all lines ME pass through a fixed point.

First Proof. As D runs along the side BC, the midpoint M of AD traces the image of the side BC in homothety $\mathcal{H}(A, \frac{1}{2})$, i.e. the midline C_1B_1. Furthermore,

$$\frac{C_1M}{MB_1} = \frac{BD}{DC} = \frac{CE}{EB},$$

so the points M and E run along segments C_1B_1 and CB in the same "relative" speed but in opposite directions.

Since $C_1B_1 \parallel CB$, there exists a negative homothety (centered at $G = BB_1 \cap CC_1$) which maps B_1C_1 to BC. From $C_1M/MB_1 = CE/EB$ we infer that such homothety also maps M to E. Hence all the lines ME pass through G.

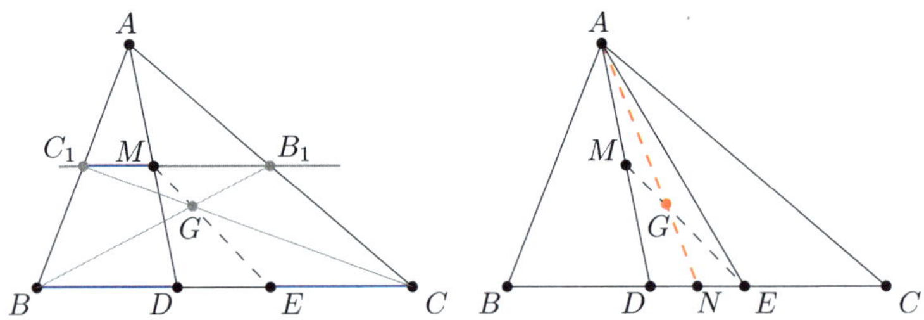

Second Proof. Let N be the common midpoint of segments DE and BC. Then the centroid G of triangle ABC is the point two-thirds of the way from A to N and hence is also the centroid of triangle ADE. Hence G lies on segment EM since it is a median of triangle ADE. Thus G is the desired fixed point.

Third Proof. Denote by N the midpoint of the side BC and by X the intersection of ME and the A-median AN. Since $ND = NE$, Menelaus' Theorem in triangle ADE for collinear points M, X, E yields

$$1 = \frac{AM}{MD} \cdot \frac{DE}{EN} \cdot \frac{NX}{XA} = 1 \cdot 2 \cdot \frac{NX}{XA}.$$

Since the ratio NX/XA does not depend on the choice of D and E, point X is the desired fixed point (note that X lies on the line ME even if $D = N$ and the triangle ADN degenerates).

8. Show that the composition of two point reflections (i.e. performing one after the other) with distinct centers O_1 and O_2 results in a translation.

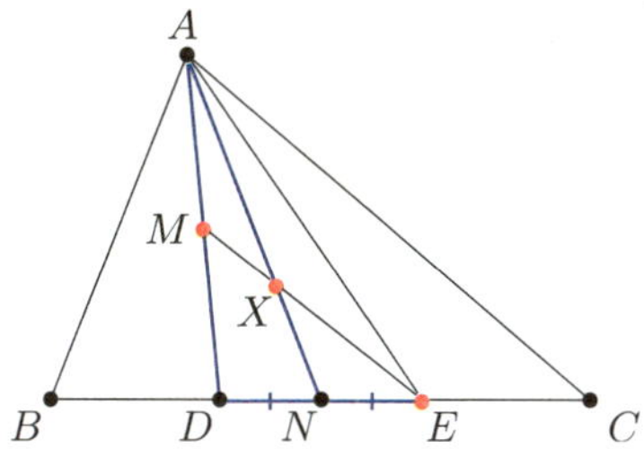

Proof. Let A be an arbitrary point, A' its reflection about O_1, and A'' the reflection of A' about O_2.

Note that O_1, O_2 are the midpoints of the segments AA', $A'A''$, respectively. If point A does not lie on the line O_1O_2, the segment O_1O_2 is a midline in triangle $AA'A''$. Hence AA'' is parallel to and twice as long as O_1O_2. In other words, point A'' is the image of A in translation by $2 \cdot \overline{O_1O_2}$.

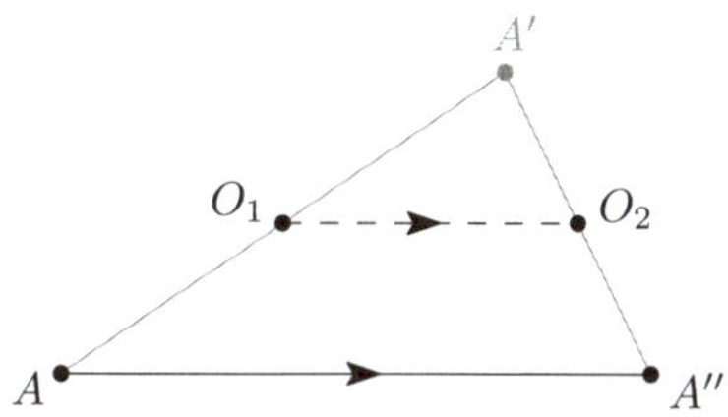

The less interesting case when A lies on the line O_1O_2 is treated using directed segments. Details are left to the reader.

9. In acute triangle ABC let A_1, B_1, C_1 be the midpoints of the respective sides and A_0, B_0, C_0 the feet of respective altitudes. Prove that the length of the closed broken line $A_0B_1C_0A_1B_0C_1A_0$ equals the perimeter of triangle ABC.

Proof. We draw the altitudes BB_0, CC_0, and the midpoint A_1 of the side BC only.

Since both $\angle BC_0C$ and $\angle BB_0C$ are right, points B_0 and C_0 lie on a circle with diameter BC. The center of this circle is precisely A_1, its radius equals $\frac{1}{2}BC$, and thus

$$C_0A_1 + A_1B_0 = \frac{1}{2}BC + \frac{1}{2}BC = BC.$$

Likewise we learn $A_0B_1 + B_1C_0 = CA$ and $B_0C_1 + C_1A_0 = AB$ and the result follows.

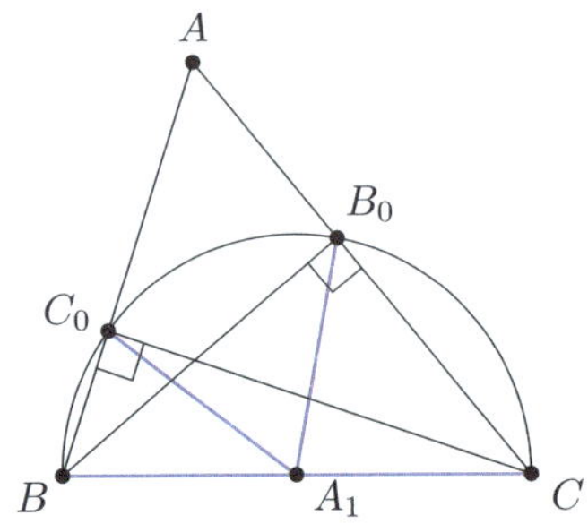

10. Fixed circles ω_1, ω_2 of distinct radii are externally tangent at T. Consider all pairs of points $A \in \omega_1$, $B \in \omega_2$ such that $\angle ATB = 90°$. Show that all such lines AB pass through a fixed point.

Proof. Let TU, TV be diameters of the circles ω_1, ω_2, respectively. Then $\angle UAT = \angle TBV = 90°$, so $UA \parallel TB$, $AT \parallel BV$, and the triangles UAT and TBV have the corresponding sides parallel. Since UT and TV have different lengths, the triangles are homothetic and thus all the lines AB passes through the center of positive homothety between UT and TV (which coincides with the center H of positive homothety between ω_1 and ω_2).

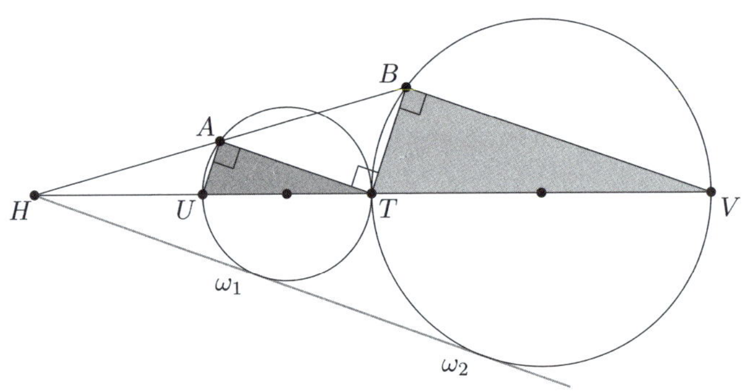

11. Let ABC be a triangle. Denote by M, N, P the midpoints of its sides BC, CA, AB, respectively, and by J, K, L the incenters of the triangles APN, BMP, CNM, respectively.

   (a) Prove that $\triangle JKL \sim \triangle ABC$.

   (b) Prove that lines JM, KN, and LP are concurrent on the line IG, where I and G are the incenter and the centroid of triangle ABC, respectively.

Proof.

(a) The midlines cut triangle ABC into four pairwise congruent triangles APN, PBM, NMC, and MNP which all have the orientation of triangle ABC. It suffices to show that triangle JKL also has this orientation.

Looking at triangles PBM and NMC we see that the segment KL connects corresponding points and thus it is equal and parallel to PN. After performing analogous arguments for other pairs of triangles we indeed learn $\triangle JKL \sim \triangle ABC$.

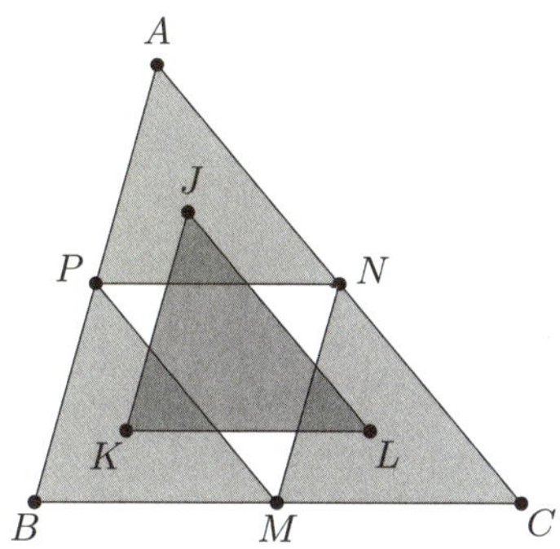

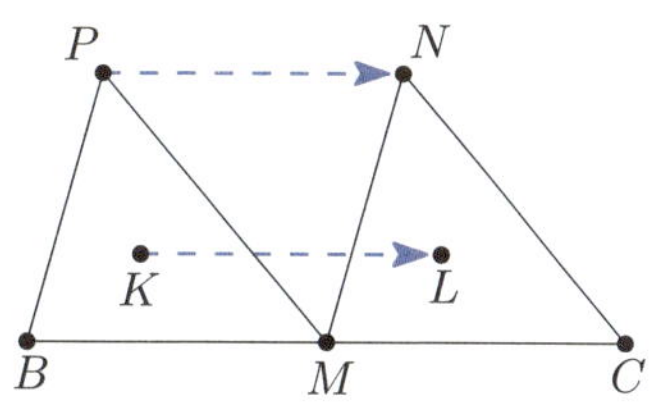

(b) From part (a) we have $\triangle JKL \sim \triangle ABC \sim \triangle MNP$ all with corresponding sides parallel. The lines are thus concurrent at the center X of homothety (which in this case is just point reflection) which takes triangle JKL to triangle MNP (see Proposition 1.28(b)).

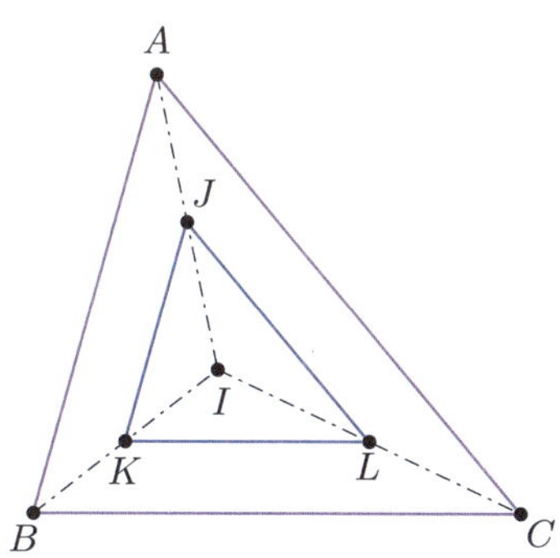

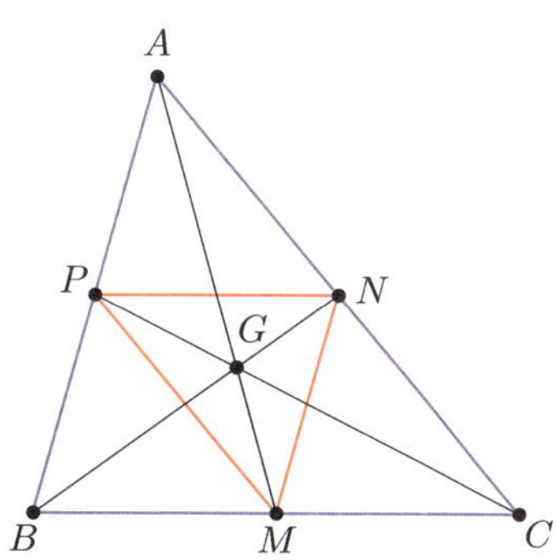

For the final part, we intend to compose homotheties. First, note that AJ, BK, CL are angle bisectors in triangle ABC and thus are concurrent at I. Therefore positive homothety which takes triangle JKL to triangle ABC is centered at I and negative homothety which takes triangle ABC to triangle MNP is centered at G (with factor $-\frac{1}{2}$). It follows that their composition is the negative homothety which sends triangle JKL to triangle MNP centered at X, hence I, G, and X are collinear (see Lemma 1.31).

12. Let ABC be a triangle with $AB < AC$. Denote by A_0 the foot of its A-altitude, by D the point of contact of the incircle with the side BC, by K the intersection of BC with the angle bisector of $\angle A$, and finally by M the midpoint of BC. Prove that points A_0, D, K, M are mutually different and lie on the line BC in this order.

Proof. Note that the points A_0, D, K, M are the projections onto BC of A, I, K, S, respectively, where I denotes the incenter of triangle ABC and S the midpoint of arc BC of its circumcircle not containing vertex A (see Proposition 1.38(b)).

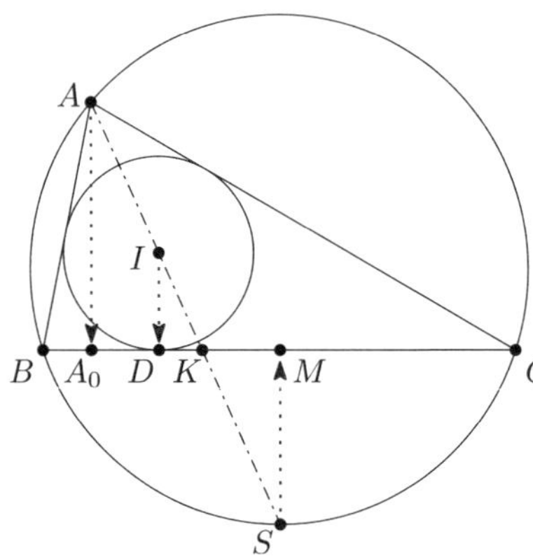

Since the points A, I, K, and S lie on the A-angle bisector in this order and are clearly mutually different, their projections are also distinct as desired, unless the A-angle bisector was perpendicular to BC. But this is obviously not the case as then AS would be the perpendicular bisector of BC and thus we would have $AB = AC$.

13. Let ω be a fixed circle with center at O and radius R and let A be a fixed point outside the circle. Point X varies on ω so that A, O, and X are not collinear. Find the locus of the intersections Y of AX with the angle bisector of $\angle AOX$,

Solution. From the Angle Bisector Theorem, we learn that

$$\frac{XY}{AY} = \frac{OX}{OA} = \frac{R}{OA},$$

which is fixed. Thus also

$$\frac{AX}{AY} = 1 + \frac{XY}{AY} = 1 + \frac{R}{OA}$$

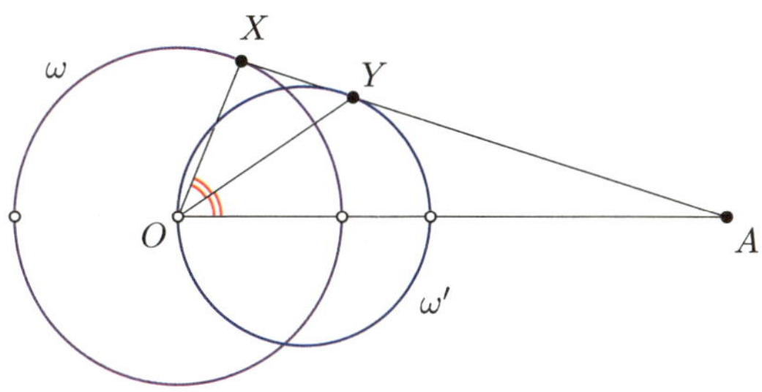

is fixed and we can say that point Y is the image of X in fixed homothety with center A and factor AY/AX. Therefore, it travels along a circle ω' which is the image of ω in this homothety attaining all admissible positions i.e. staying off the line OA.

Remark. The reader is encouraged to verify that $O \in \omega'$ although it is not part of the sought-after locus.

14. A variable point X runs along a semicircle ω with diameter AB ($X \neq A$, $X \neq B$). Let Y be such point on the ray XA that $XY = XB$. Find the locus of points Y.

Solution. Triangle XYB is isosceles and right, therefore Y is the image of X in spiral similarity $\mathcal{S}(B, \sqrt{2}, +45°)$. The locus is thus the image of ω (excluding points A and B) in this spiral similarity. To be more specific, it is the semicircle (without its endpoints) with one endpoint at B and the midpoint at A.

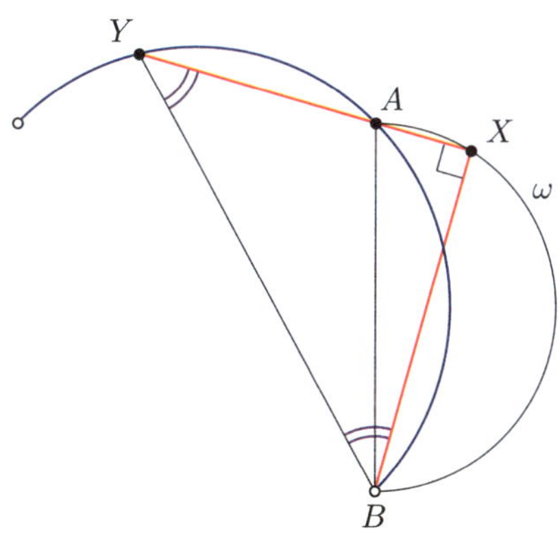

15. A variable regular hexagon $ABCDEF$ has fixed point A and its center O is moving along a given line. Prove that the remaining five vertices also describe straight lines and that these lines are concurrent.

Proof. As the shape of $ABCDEF$ is fixed, points B, C, D, E, and F are images of O in fixed spiral similarities (possibly degenerate into rotations or homotheties) centered at A. For example $\mathcal{S}(A, \frac{AE}{AO}, \angle(OA, AE))$ (which can be simplified as $\mathcal{S}(A, \sqrt{3}, +30°)$) sends O to E, and the others would be found similarly. Therefore the remaining five vertices indeed describe straight lines.

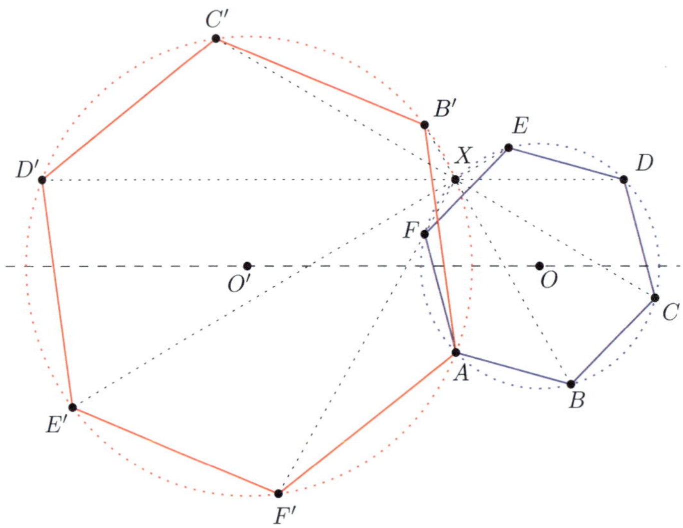

Now consider two positions of the hexagon $ABCDEF$ with center O and $AB'C'D'E'F'$ with center O'. Being familiar with spiral similarity, we recall that lines BB', CC', DD', EE' and FF' all pass through the second intersection X of the circumcircles of $ABCDEF$ and $AB'C'D'E'F'$ (see Proposition 1.48(a)). But since both circles are symmetric with respect to line OO', point X is just a reflection of A over this line and therefore is independent of the choice of hexagons.

16. Let $ABCD$ be a cyclic quadrilateral and let H_d, H_c be the orthocenters of the triangles ABC and ABD, respectively.

   (a) Show that points A, B, H_d, H_c lie on a single circle.

   (b) Draw also H_a and H_b, the orthocenters of triangles BCD and CDA, and prove that $ABCD$ is congruent to $H_aH_bH_cH_d$.

Proof.

(a) The images H'_d and H'_c of H_d and H_c under reflection in line AB lie on the circumcircle ω of $ABCD$ (see Proposition 1.36). But then the image ω' of ω in the same reflection contains points A, B, H_d, and H_c so they are apparently concyclic.

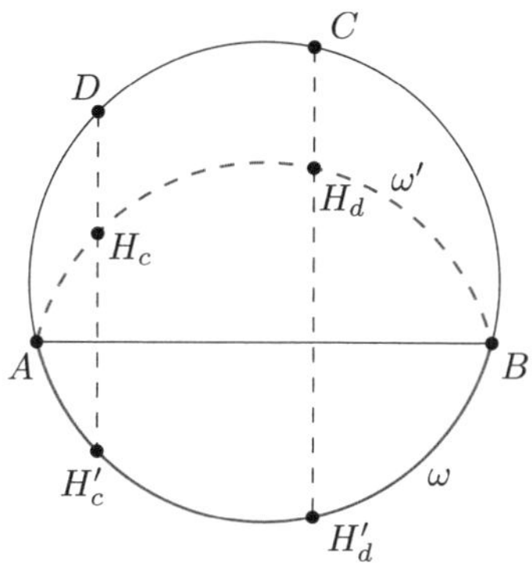

(b) We shall only prove that $CD \parallel H_dH_c$ and $CD = H_dH_c$, since if in two quadrilaterals the corresponding sides are parallel and equal, the quadrilaterals are congruent.

We work again with the reflections H'_d and H'_c and focus on the strip between parallel lines DH_c and CH_d.

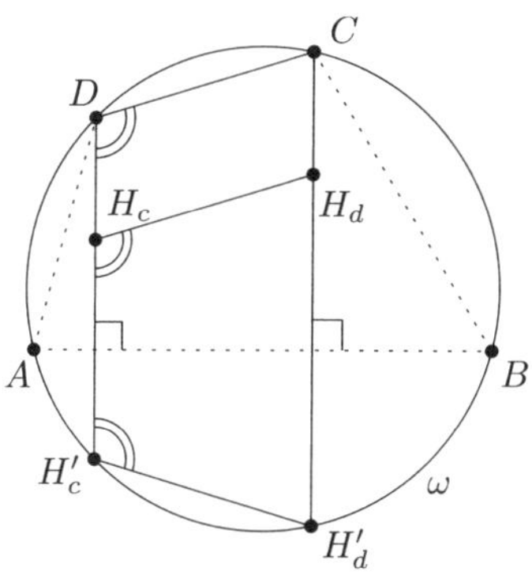

Observe that both DC and H_cH_d are reflections of $H'_cH'_d$ across a line parallel to AB (which in the first case is a diameter of ω parallel to AB). Therefore, they are equal and as they are both antiparallel with $H'_cH'_d$ with respect to line AB, they are parallel themselves. We are done.

17. [China Girls 2012] Let D and E be the points of contact of the incircle of triangle ABC with its sides AB and AC, respectively. Also, let X be the circumcenter of triangle BIC, where I is the incenter of triangle ABC. Show that $\angle XDB = \angle XEC$.

    Proof. Recall that the circumcenter of BIC is the midpoint of arc BC of the circumcircle of triangle ABC (see Proposition 1.38(b)). In particular, it lies on AI so let us draw it vertically. As $AD = AE$, the quadrilateral $ADXE$ is symmetric about AI and the conclusion follows since $\angle XDB$ and $\angle XEC$ correspond in this symmetry.

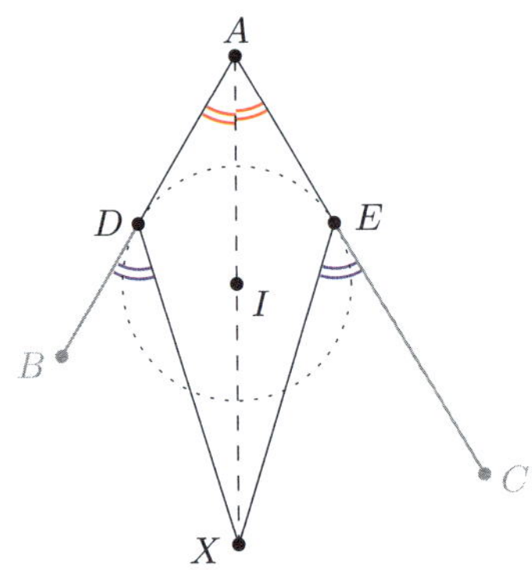

18. Let ABC be a scalene acute-angled triangle with orthocenter H. Show that the Euler lines[1] of triangles BHC, CHA, AHB intersect at one point on the Euler line of triangle ABC.

    First Proof. We look at triangle BHC and recall that its orthocenter is A and that its circumcircle is symmetric with the one of triangle ABC (see Proposition 1.35(d)), therefore the circumcenter O' of triangle BHC is the reflection of O (the circumcenter of triangle ABC) across BC.

    We will prove that AO' intersects the Euler line OH of triangle ABC at a fixed point. Observe that if we denote by M the midpoint of BC, then AM is a common median of triangles ABC and AOO' and so their centroids coincide at point G. But then the midpoint X of AO' lies on OG and $2 \cdot GX = GO$ (centroid divides the median in ratio $2:1$). Hence all four Euler lines pass through X.

    Second Proof. Take a good look at the nine-point circles (see Theorem 1.37) of triangles BHC, CHA, AHB and observe that they in fact all coincide with the nine-point circle of triangle ABC (if in trouble see also

[1]For explanation see Example 1.3

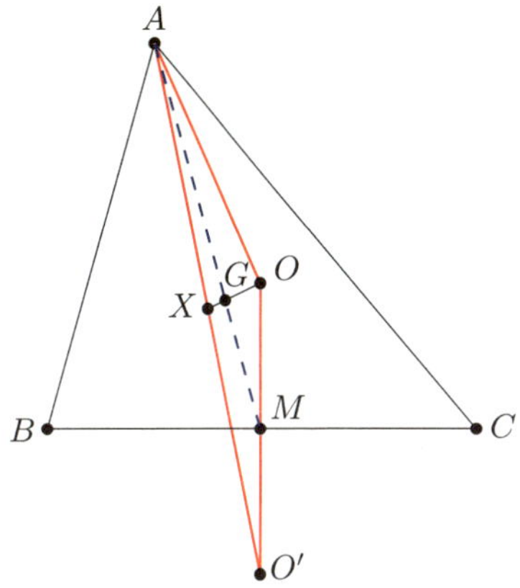

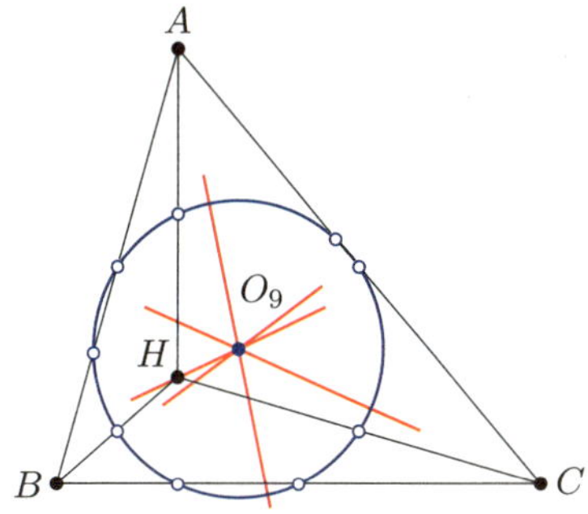

Lemma 1.34). Thus, all four Euler lines pass through the common center O_9.

19. [based on IMO shortlist 2011] Let ABC be a triangle and D the foot of its A-altitude. The line through A parallel to BC intersects the circumcircle ω of triangle ABC for the second time at E. Prove that line DE passes through the centroid of triangle ABC.

Proof. Denote by M the midpoint of BC and by X the intersection of AM and DE. It suffices to prove that $MX : XA = 1 : 2$.

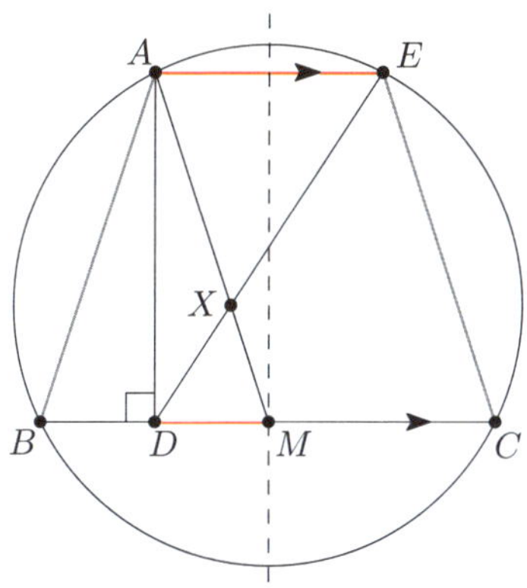

From similar triangles MXD and AXE we have

$$\frac{MX}{XA} = \frac{DM}{AE}$$

where the latter indeed equals $\frac{1}{2}$, since the cyclic trapezoid $BCEA$ is isosceles and therefore symmetric with respect to the perpendicular bisector of BC.

20. [Putnam 1996] Let ω_1 and ω_2 be circles whose centers O_1, O_2 are 10 units apart and whose radii are 1 and 3 units. Find the locus of points M which are the midpoints of some segment XY, where $X \in \omega_1$ and $Y \in \omega_2$.

Solution. First, fix a point Y on ω_2. The midpoints of XY, where $X \in \omega_1$, form a circle which is the image of ω_1 in homothety $\mathcal{H}(Y, \frac{1}{2})$. Therefore its radius is $\frac{1}{2}$ and its center is the midpoint of YO_1.

Now as Y varies, the midpoints of YO_1 move along a circle ω_2' which is the image of ω_2 in homothety $\mathcal{H}'(O_1, \frac{1}{2})$. The radius of ω_2' is thus $\frac{3}{2}$ and its center is the midpoint of O_1O_2.

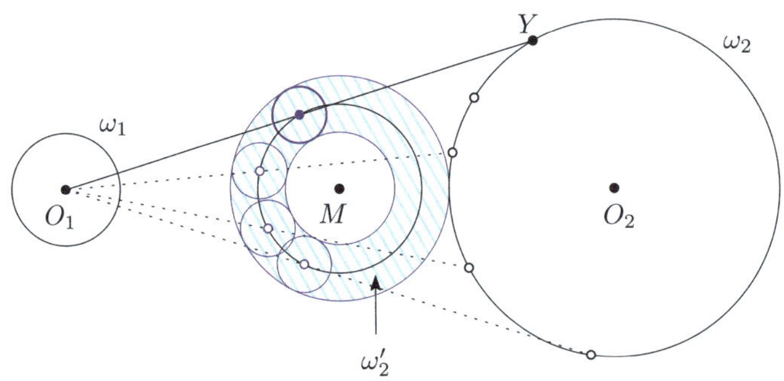

Altogether, we see that the locus of all possible midpoints of XY is annular region centered at the midpoint M of O_1O_2 with inner radius $\frac{3}{2} - \frac{1}{2} = 1$ and outer radius $\frac{3}{2} + \frac{1}{2} = 2$.

21. [USAMTS 2005] Let ω be a given circle. Points A, B, and C lie on ω such that ABC is an acute triangle. Points X, Y, and Z are also on ω such that $AX \perp BC$ at D, $BY \perp AC$ at E, and $CZ \perp AB$ at F. Show that the value of

$$\frac{AX}{AD} + \frac{BY}{BE} + \frac{CZ}{CF}$$

does not depend on the choice of A, B and C.

Proof. The lines AX, BY, and CZ are altitudes in triangle ABC which intersect at its orthocenter H.

Moreover, X, Y, and Z are the images of H under reflections about BC, CA, AB, respectively (see Proposition 1.36) and we can rewrite the ratios to ratios of areas as follows:

$$\frac{AX}{AD} = 1 + \frac{DX}{AD} = 1 + \frac{DH}{DA} = 1 + \frac{[BHC]}{[ABC]}.$$

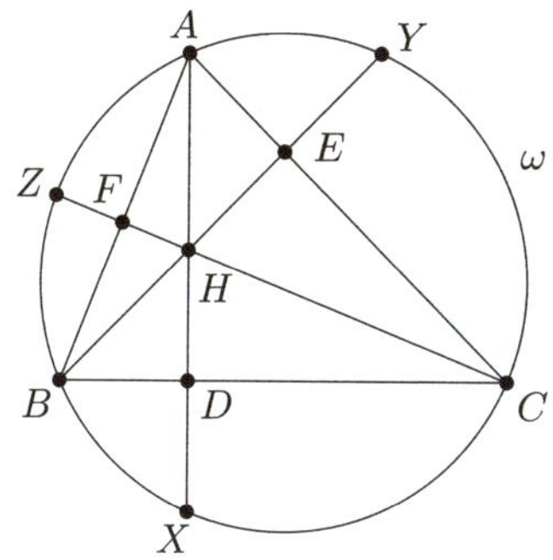

Finally, by analogy we see that

$$\frac{AX}{AD} + \frac{BY}{BE} + \frac{CZ}{CF} = 3 + \frac{[BHC] + [CHA] + [AHB]}{[ABC]} = 4,$$

where in the last equality we have used that H lies inside (acute) triangle ABC.

22. Let ABC be a triangle with $\angle A = 90°$ and let L be a point on BC. The circumcircles of the triangles ABL and ACL intersect AC and AB for the second time at M and N, respectively. Prove that $BM \perp CN$.

First Proof. Lines BM and CN do not have much in common but thanks to two cyclic quadrilaterals they both form a convenient angle with BC. There are more configurations possible but either way the angle-chasing

$$\angle MBC + \angle BCN = \angle CAL + \angle BAL = 90°$$

implies that $BM \perp CN$ as desired.

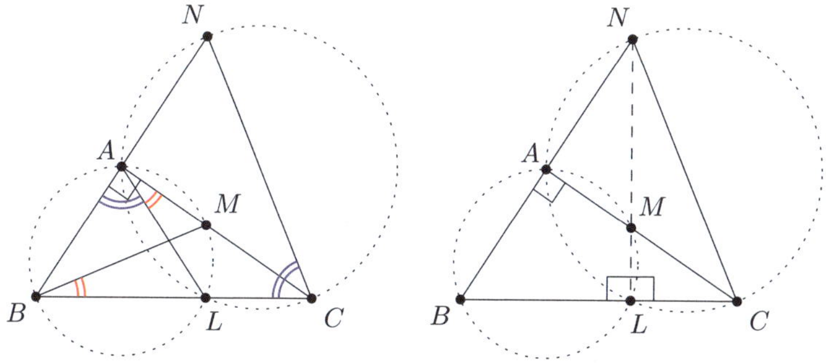

Second Proof. Given a right angle and circles, there are always more right angles hidden. In our case $\angle BLM = \angle BAM = 90°$ and $\angle CLN = \angle CAN = 90°$. Hence the points L, M, N are collinear and $NL \perp BC$.

Now what is M with respect to triangle NBC? It is the intersection of two altitudes (namely CA and NL), so it is the orthocenter and $BM \perp CN$ too.

23. Triangle centers in other roles.

Let ABC be an acute triangle. *Pedal triangle* of a point X is the triangle formed by the projections of X onto the triangle sides. Denote by I, O, H the incenter, circumcenter, and orthocenter of triangle ABC, respectively.

(a) Prove that I is the circumcenter of its pedal triangle.

(b) Prove that O is the orthocenter of its pedal triangle.

(c) Prove that H is the incenter of its pedal triangle.

Proof.

(a) The projections of I onto the triangle sides are simply the points of contact of the incircle. Since I is the center of the incircle, the result follows.

(b) The projections of O onto the triangle sides BC, CA, AB are their midpoints A_1, B_1, C_1. Since midline is parallel to the base, the perpendicular bisector of BC coincides with the A_1-altitude of triangle $A_1B_1C_1$. We conclude by applying this idea cyclically.

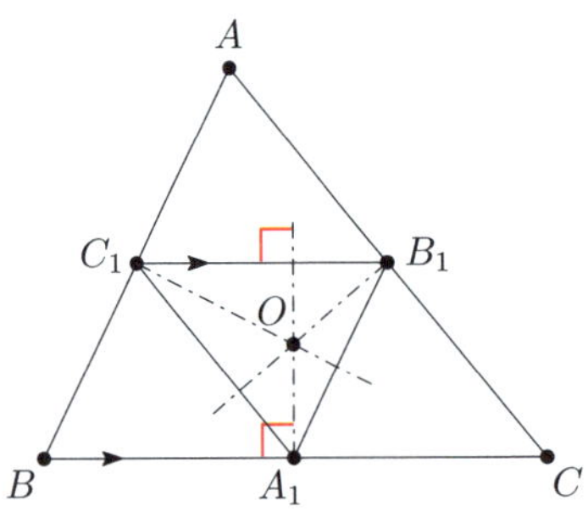

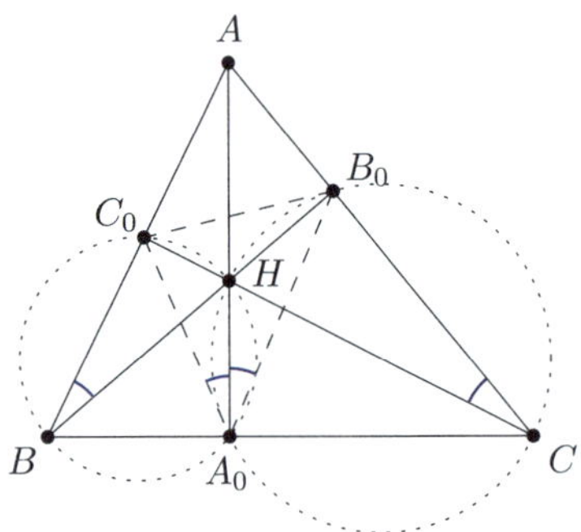

(c) The projections of H onto the triangle sides are the respective feet of altitudes A_0, B_0, C_0.

We will prove that A_0A is the angle bisector in triangle $A_0B_0C_0$. Recall that quadrilaterals BA_0HC_0, CA_0HB_0, and BCB_0C_0 are cyclic (see Proposition 1.35(a),(b)). It follows that

$$\angle AA_0C_0 = \angle HBC_0 = \angle B_0CH = \angle B_0A_0A.$$

Likewise we show that BB_0 and CC_0 are also angle bisectors in triangle $A_0B_0C_0$ and thus H is indeed the incenter of triangle $A_0B_0C_0$.

24. Given a right triangle ABC, let $ABDE$ be a square erected outwards from its hypotenuse AB. Prove that the angle bisector of $\angle C$ bisects the area of the square $ABDE$.

First Proof. First, what lines bisect the area of a given square? Since square is centrally symmetric, these are precisely the lines that pass through its center. Hence instead of dealing with D and E, let O be the center of $ABDE$, i.e. the third vertex of right O-isosceles triangle ABO erected outwards from AB. Now it suffices to prove that CO bisects angle ACB.

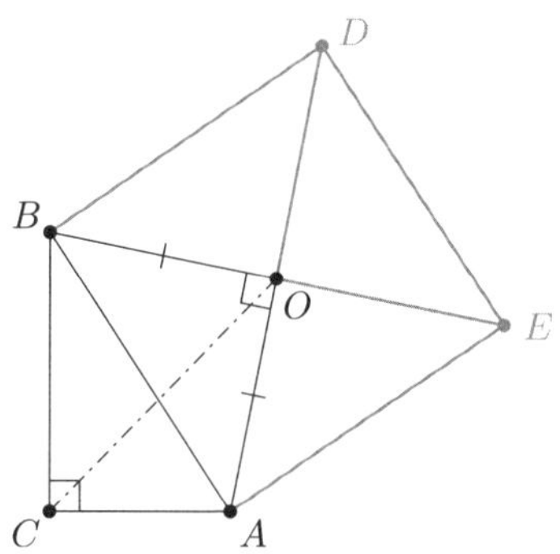

But this is readily done, as $\angle AOB = \angle ACB = 90°$ and $OA = OB$ imply that O is the midpoint of arc AB of the circumcircle of triangle ABC not containing vertex C, and as such it lies on the angle bisector of $\angle C$ (see Proposition 1.38(b)).

Second Proof. Build a square $CXYZ$ circumscribed about $ABDE$ by adding right triangles BXD, DYE, and EZA congruent to triangle ABC. Then the angle bisector of $\angle C$ is clearly CY. It passes through the common center O of $ABDE$ and $CXYZ$, hence it bisects the area of the square $ABDE$.

25. [Tournament of Towns 2010] Let $ABCD$ be a rhombus with a point P on the side BC and Q on the side CD such that $BP = CQ$. Prove that the centroid of the triangle APQ lies on the segment BD.

Proof. Since the centroid is usually difficult to handle, we first try to restate the problem. Recalling that the centroid "trisects" the median,

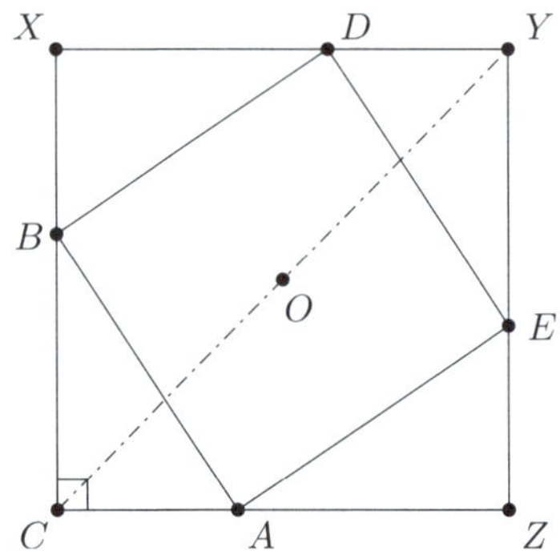

the statement equivalently says that the midpoint of PQ lies on the image of line BD in homothety $\mathcal{H}(A, \frac{3}{2})$, which is the midline EF (with $E \in BC$, $F \in CD$) in isosceles triangle DBC. Now if we note that $BP = CQ$ rewrites as $EP = FQ$, we may conveniently forget more than half of the picture.

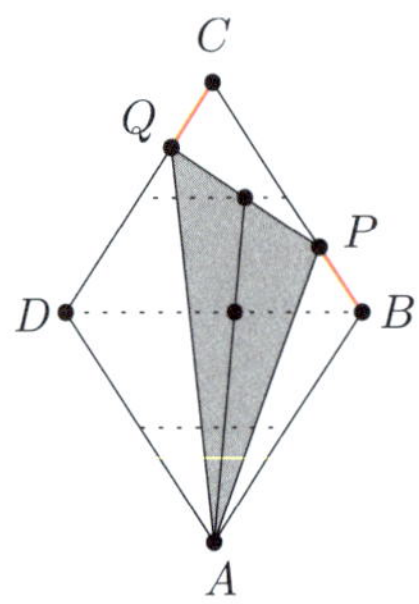

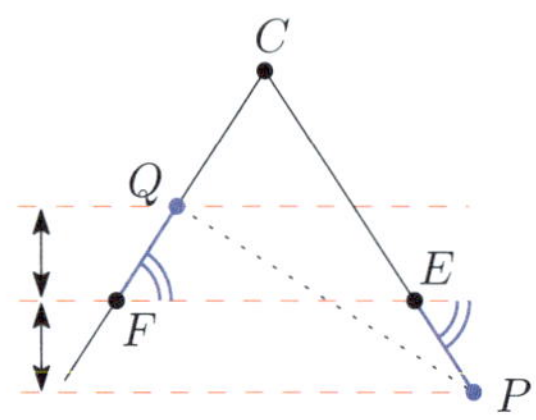

Proving that EF bisects PQ is not difficult. Since $EP = FQ$ and the lines CE and CF subtend the same angle with EF, points P and Q have the same distance from the line EF. As they lie in the opposite half-planes, the midpoint of PQ lies on EF as desired.

26. [Romania 2006] Let ABC be a triangle. Points M, N on its sides AB, AC, respectively, satisfy

$$\frac{BM}{AB} = 2 \cdot \frac{CN}{AC}.$$

The line perpendicular to MN passing through N intersects side BC at P. Prove that $\angle MPN = \angle NPC$.

Proof. Place BC horizontally. The given condition then states that the point M is twice as high above BC as N. In other words, if ℓ is a line through M parallel to BC then the point N lies midway between BC and ℓ. Denoting the intersection of the lines MN and BC by L we conclude that N is the midpoint of ML.

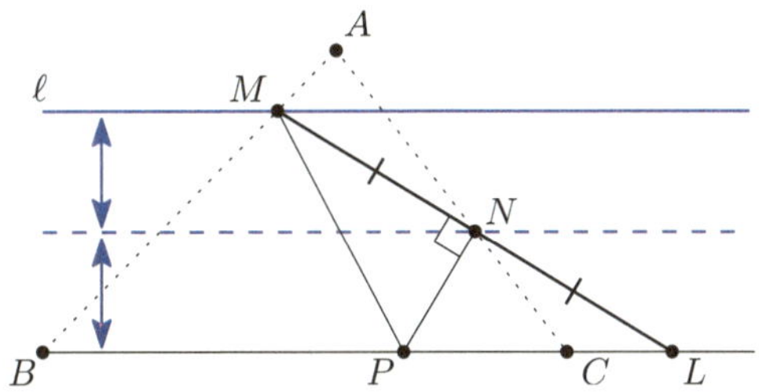

Thus in triangle MPL both the P-median and P-altitude coincide with PN implying that triangle MPL is isosceles (if in doubt, consult Introductory Problem 12). Hence PN is simultaneously the angle bisector.

27. Let ABC be a scalene triangle and denote by D the intersection of the external angle bisector at A with line BC. Prove that

   (a) $DB/DC = AB/AC$.

   (b) If we define points $E \in AC$ and $F \in AB$ also as feet of the respective external angle bisectors, then D, E, and F are collinear.

Proof.

(a) Denote the external angle bisector by ℓ and place it horizontally. Now we see that both DB/DC and AB/AC express the ratio of distances of the points B and C to the line ℓ.

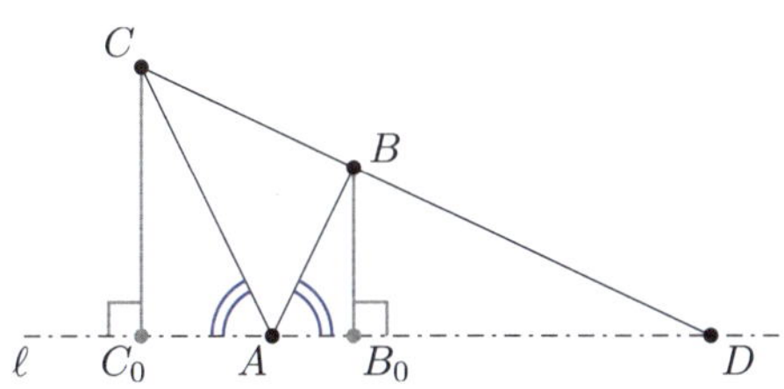

Indeed, let B_0, C_0 be the projections of B and C onto ℓ, respectively. Then $\triangle DBB_0 \sim \triangle DCC_0$ and $\triangle ABB_0 \sim \triangle ACC_0$ (AA), hence

$$\frac{DB}{DC} = \frac{BB_0}{CC_0} = \frac{AB}{AC}.$$

Remark. This result is known as the External Angle Bisector Theorem and in its full form it also asserts the converse: If a point on the line BC (but outside the segment BC) satisfies the equality of ratios, it must be the foot of the external angle bisector. The proof we gave can be easily adapted to cover this converse statement. Try it!

(b) By Menelaus' Theorem, points D, E, F are collinear if and only if

$$\frac{BD}{DC} \cdot \frac{CE}{EA} \cdot \frac{AF}{FB} = 1.$$

Using part (a) we can replace each of the ratios on the left-hand side and rewrite the latter (equivalently) as

$$\frac{BA}{AC} \cdot \frac{CB}{BA} \cdot \frac{AC}{CB} = 1,$$

which is true. Problem solved.

28. Let ABC be a scalene acute triangle. Draw points K, L, M, N such that $ABMN$ and $LBCK$ are congruent rectangles erected outwards from the triangle sides. Prove that lines AL, NK, MC are concurrent.

First Proof. Denote the intersection of AL and CM by X. For the rectangles to be congruent we must have $MB = BC$ and $AB = BL$, therefore the triangles MBC and ABL are both isosceles. As $\angle MBC = 90° + \angle B = \angle ABL$ we even have $\triangle MBC \sim \triangle ABL$ and $\angle XMB = \angle XAB$. Thus, X lies on the circumcircle of $ABMN$ and similarly on the circumcircle of $LBCK$.

Now as BN and BK are diameters it follows that $\angle BXN = 90°$ and $\angle KXB = 90°$ implying that points N, X, and K are collinear.

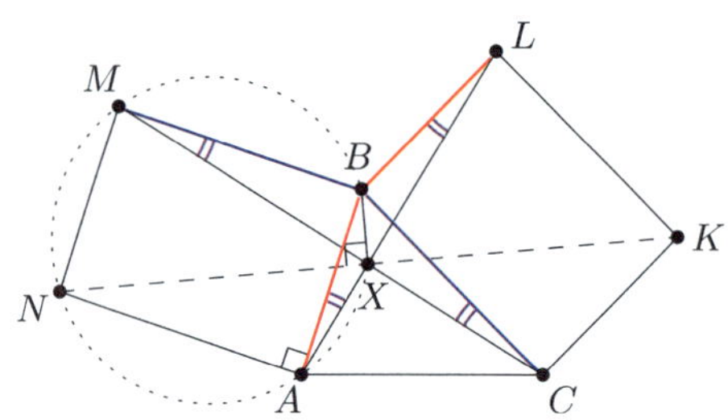

Second Proof. Consider rotation centered at B carrying $BMNA$ to $BCKL$. As rotation is a special case of spiral similarity, the Proposition 1.48 implies that all three lines pass through the second intersection of the circumcircles of the rectangles $ABMN$ and $LBCK$.

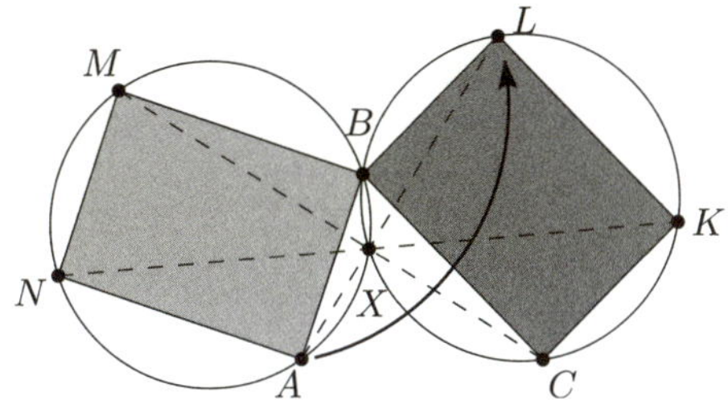

29. [USAMO 1993] Let $ABCD$ be a convex quadrilateral whose diagonals intersect at right angle at O. Prove that the reflections of O across lines AB, BC, CD, DA are concyclic.

First Proof. Instead of reflections across the sides of $ABCD$ we shall work with the projections A', B', C', and D' of O on AB, BC, CD, DA, respectively. Once we prove A', B', C', and D' are concyclic, the conclusion will follow from the homothety $\mathcal{H}(O,2)$.

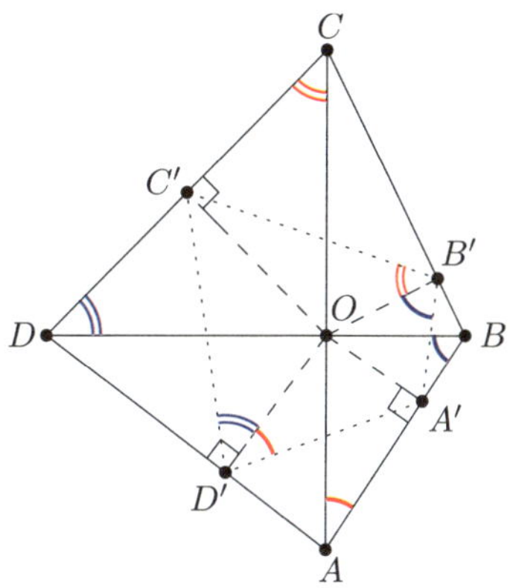

Observe that quadrilaterals $A'BB'O$, $B'CC'O$, $C'DD'O$, and $D'AA'O$ are cyclic with diameters BO, CO, DO, and AO, respectively. We will use this to show $\angle A'D'C' + \angle C'B'A' = 180^\circ$. Indeed, we have

$$\angle A'D'C' = \angle A'D'O + \angle OD'C' = \angle BAO + \angle ODC,$$

and similarly

$$\angle C'B'A' = \angle C'B'O + \angle OB'A' = \angle DCO + \angle OBA,$$

but looking at the right triangles DOC and AOB we see that the sum of these angles is 180°.

Second Proof. As in the first proof we note that it suffices to prove that the points A', B', C', D' are concyclic.

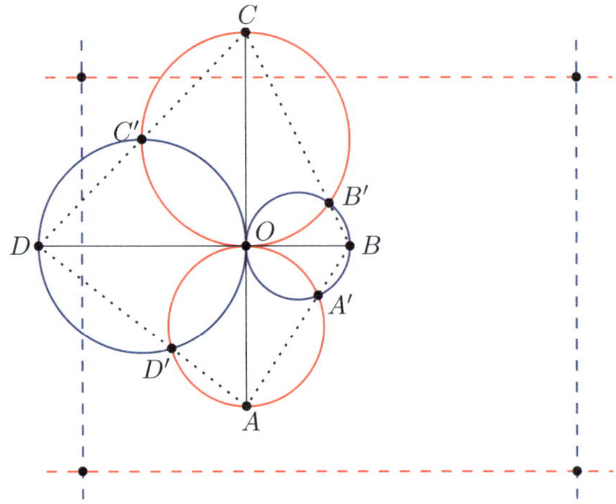

Draw the diagram so that DB is horizontal and AC is vertical. We invert about O.

The lines BD and AC will remain horizontal and vertical, respectively. The circumcircle of $OA'BB'$ which has diameter OB and the circumcircle of $OC'DD'$ which has diameter OD will become vertical lines. Likewise, the circumcircles of $OD'AA'$ and $OB'CC'$ will become horizontal lines. Hence $A'B'C'D'$ will become a rectangle. Since the images of A', B', C', D' lie on a circle not passing through O, so do the original points A', B', C', and D'.

30. Let $ABCD$ be a cyclic quadrilateral and let I_1, I_2 be the incenters of the triangles ABC and ABD, respectively.

    (a) Show that the quadrilateral ABI_1I_2 is cyclic.

    (b) Draw also I_3 and I_4, the incenters of triangles CDA and BCD, and prove that $I_1I_2I_3I_4$ is a rectangle.

Proof.

(a) It suffices to show that $\angle AI_1B = \angle AI_2B$. We have (recalling Proposition 1.38(a))

$$\angle AI_1B = 90^\circ + \frac{1}{2}\angle ACB, \quad \angle AI_2B = 90^\circ + \frac{1}{2}\angle ADB$$

but since $ABCD$ is cyclic it follows that $\angle ACB = \angle ADB$ and we are done.

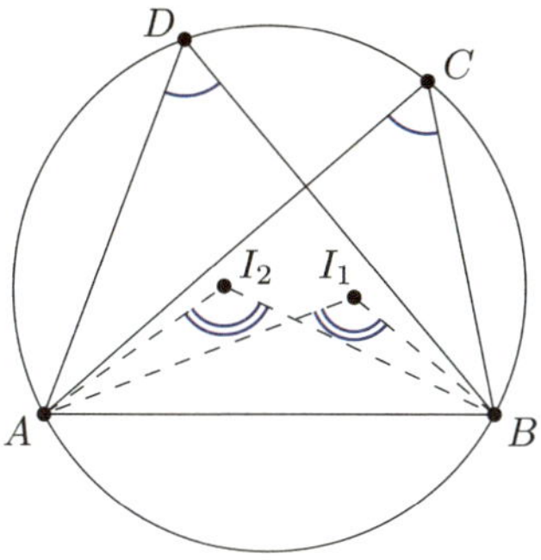

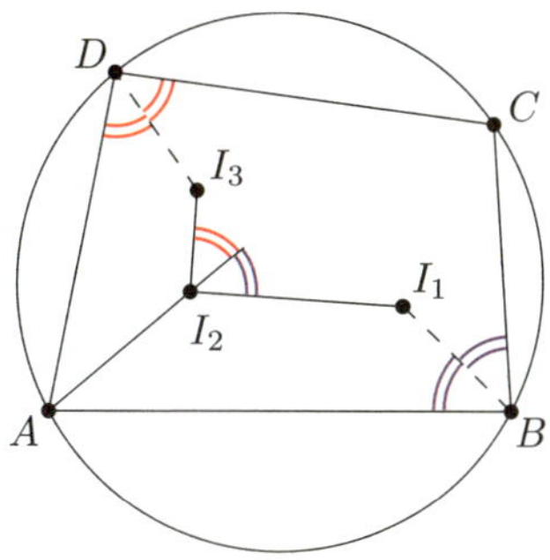

(b) (Japanese theorem for cyclic quadrilaterals) We will show that $I_1I_2 \perp I_2I_3$. From (a) we know that ABI_1I_2 and ADI_3I_2 are cyclic. Extending the ray AI_2 beyond I_2, we see that

$$\angle I_1I_2I_3 = \frac{1}{2}\angle ABC + \frac{1}{2}\angle CDA = 90^\circ.$$

Similarly, we show $I_2I_3 \perp I_3I_4$ and $I_3I_4 \perp I_4I_1$ which completes the proof.

31. Let M be the midpoint of the side BC of a triangle ABC. Point K on the segment AM satisfies $CK = AB$. Denote by L the intersection of CK and AB. Prove that triangle AKL is isosceles.

First Proof. In order to "connect" equal segments AB and CK and make use of the midpoint M of BC, let K' be the point such that $BKCK'$ is a parallelogram. Then K' lies on the A-median (beyond M) and $K'B = CK = AB$. Hence triangle ABK' is B-isosceles and since $K'B$ and CK are parallel, triangle AKL is L-isosceles.

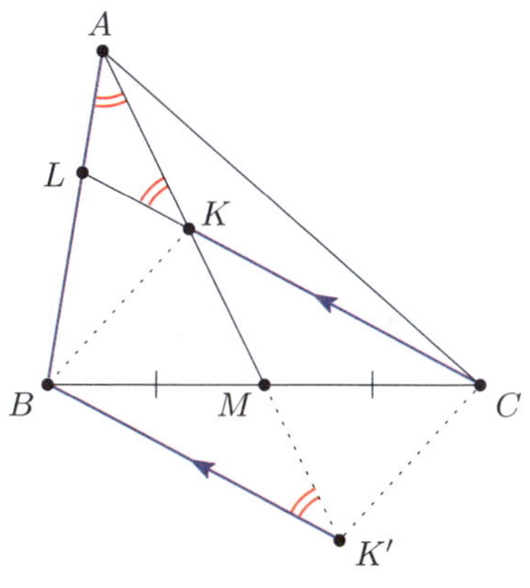

Second Proof. This time we exploit equal segments $CK = AB$ and $BM = MC$ by means of Menelaus' Theorem in triangle LBC for collinear points A, K, M. We obtain

$$1 = \frac{LA}{AB} \cdot \frac{BM}{MC} \cdot \frac{CK}{KL} = \frac{LA}{KL}.$$

Hence triangle AKL is L-isosceles.

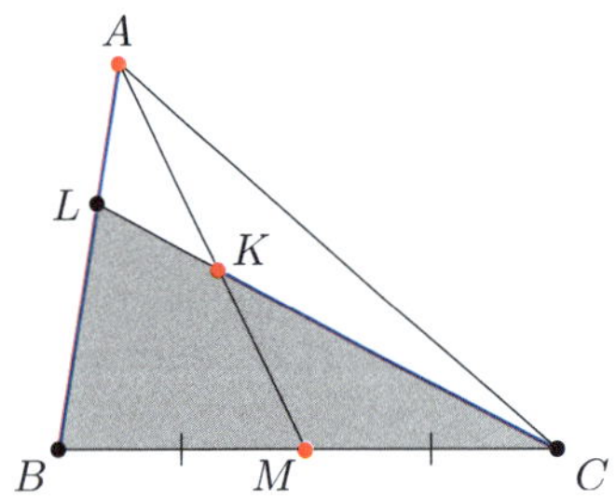

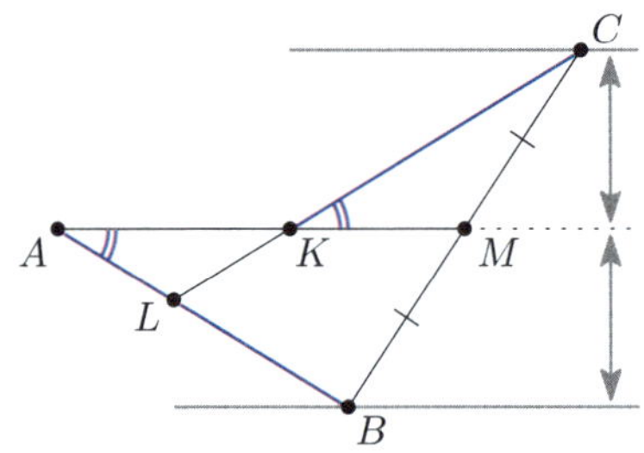

Third Proof. Place AM horizontally. As M is the midpoint of BC, point C is as much "above" AM as B is "below" it. Since the segments AB and CK are equal, they form the same angle with AM. Thus, triangle ALK is isosceles.

32. Let A_1, B_1, C_1 be the midpoints of the arcs BC, CA, AB of the circumcircle of triangle ABC (not containing A, B, C, respectively) and let A_2, B_2, C_2 be the tangency points of the incircle with BC, CA, AB, respectively. Prove that the lines A_1A_2, B_1B_2, C_1C_2 are concurrent.

Proof. Place BC horizontally with A "above" it and observe that A_1 and A_2 are both the "bottom" points on the respective circles.

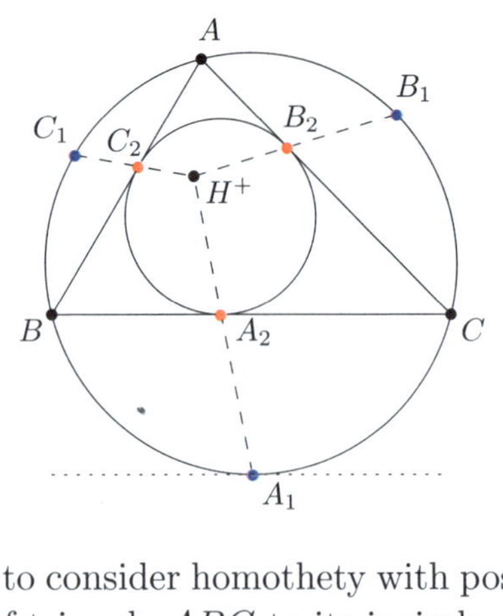

Thus it is natural to consider homothety with positive factor which takes the circumcircle of triangle ABC to its incircle.

As A_1 and A_2 correspond in this homothety, line A_1A_2 passes through its center H^+. For analogous reasons also B_1B_2 and C_1C_2 pass through H^+ and the concurrence is proved.

33. Let ABC be a triangle with incenter I and A-excenter E. Further, let M be the midpoint of arc BC that does not contain A, and let $D = AI \cap BC$. Prove the following metric identities:

    (a) $AD \cdot AM = AB \cdot AC$.

    (b) $AI \cdot AE = AB \cdot AC$.

    (c) $MA \cdot ID = MI \cdot AI$.

First Proof.

(a) Observe that $\angle AMB = \angle ACB$ and thus $\triangle ABM \sim \triangle ADC$ (AA). From this similarity we get

$$\frac{AB}{AM} = \frac{AD}{AC},$$

and the result follows.

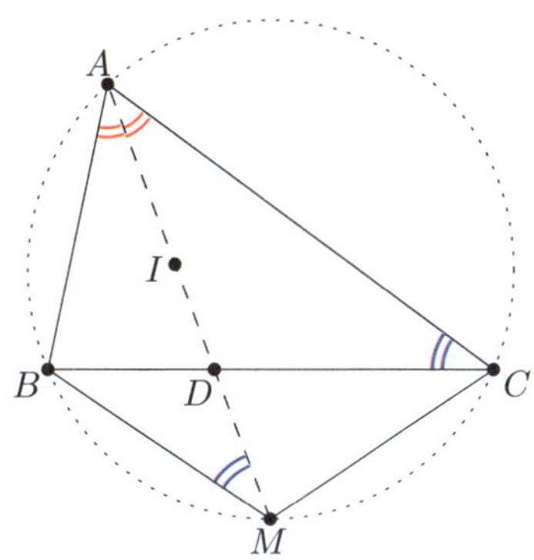

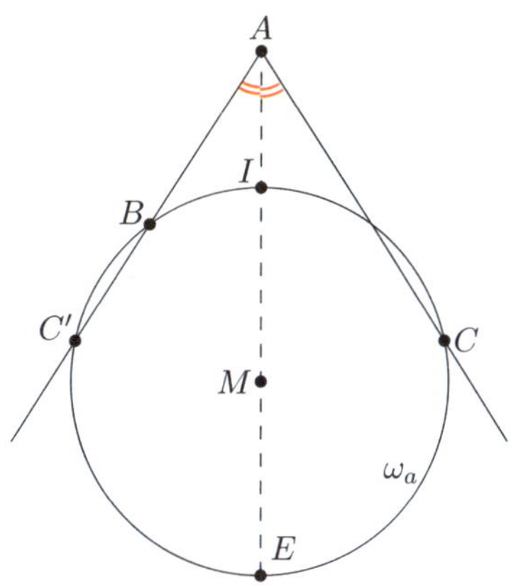

(b) Place AI vertically. We recall that points B, I, C, E lie on a circle centered at M (see the Big Picture, Proposition 1.42(b)) and call the circle ω_a. Aiming to use Power of a Point, further, we denote by C' the second intersection of AB and ω_a and learn

$$AI \cdot AE = AB \cdot AC',$$

but from symmetry in line AI, we have $AC' = AC$ and we may conclude.

(c) We write $ID = MI - MD$. Thus, using the Shooting Lemma (see Proposition 1.40(b)), we obtain

$$MA \cdot ID = MA \cdot MI - MA \cdot MD = MA \cdot MI - MI^2$$
$$= MI \cdot (MA - MI) = MI \cdot AI.$$

Second Proof. Parts (a) and (b) follow also from $\sqrt{bc}$-inversion. The fact that the image of D is M ensures part (a) and for part (b) we remark (in the interesting case of a scalene triangle) that the image of the circumcircle of $BICE$ is a circle centered somewhere on AI and passing through B and C. Hence it is its own image and I maps to E.

34. Points M and N vary over the interiors of the sides AB and AC of a triangle ABC so that $BM/MA = AN/NC$. Prove that the circumcircles of the triangles AMN pass through another fixed point different from A.

Proof. Let S be the center of spiral similarity that maps segment BA to AC (in this order of vertices), namely $\mathcal{S}(S, AC/AB, \angle(BA, AC))$. Since the points M, N divide the corresponding segments BA, AC in the same ratio, similarity $\mathcal{S}$ also maps M to N, implying that $\angle(MS, SN) = \angle(BA, AC)$. Quadrilateral $AMSN$ is then cyclic and the conclusion follows.

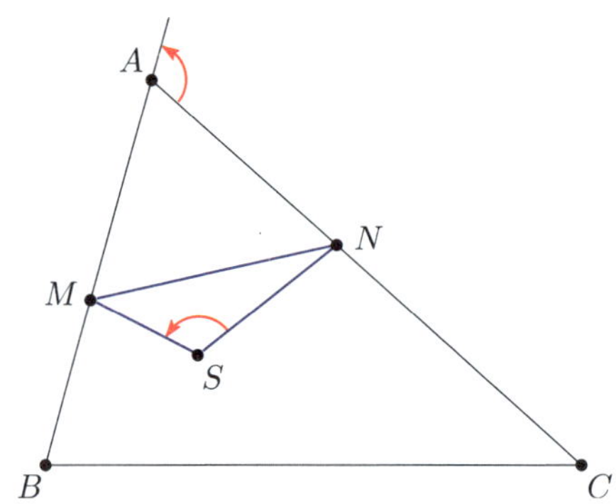

35. [Romania 2001] A triangle ABC and a point D in its interior are given. Consider points E, F such that $\triangle AFB \sim \triangle CEA \sim \triangle CDB$, points B and E lie on different sides of the line AC, and points C and F lie on different sides of AB. Prove that $AEDF$ is a parallelogram.

Proof. Denote $\angle(CE, CA)$ by φ and CA/CE by k. Then spiral similarity $\mathcal{S}(C, k, \varphi)$ takes E to A and D to B. Therefore it takes ED to AB and so $\angle(ED, AB) = \varphi$. Since also $\angle(AF, AB) = \varphi$, we have $ED \parallel AF$. Likewise we get $FD \parallel AE$ which ensures that $AEDF$ is a parallelogram.

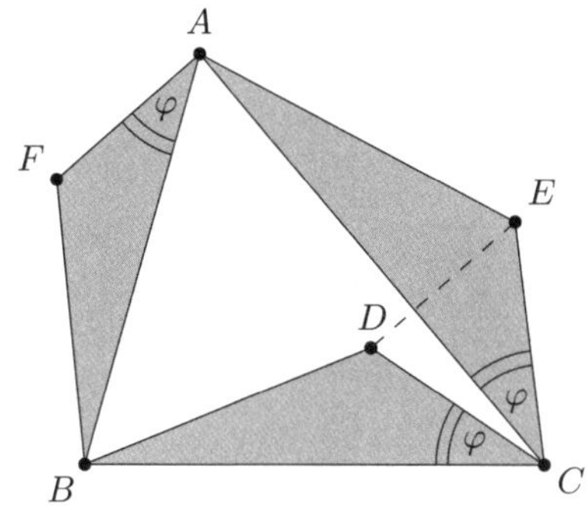

36. Napoleon[2]'s Theorem

Let ABC be a triangle and let BCD, CAE, ABF be equilateral triangles erected outwards from its sides. Show that the centroids A_1, B_1, C_1 of these equilateral triangles also form an equilateral triangle.

First Proof. Spiral similarity $\mathcal{S}(C, \sqrt{3}, +30^\circ)$ takes B_1 to A and A_1 to D. Thus it takes B_1A_1 to AD and so $AD = \sqrt{3} \cdot B_1A_1$. The same argument with spiral similarity $\mathcal{S}'(B, \sqrt{3}, -30^\circ)$ shows that also $AD = \sqrt{3} \cdot C_1A_1$. Therefore we have $B_1A_1 = C_1A_1$ and likewise we obtain $B_1A_1 = B_1C_1$, which ends the proof.

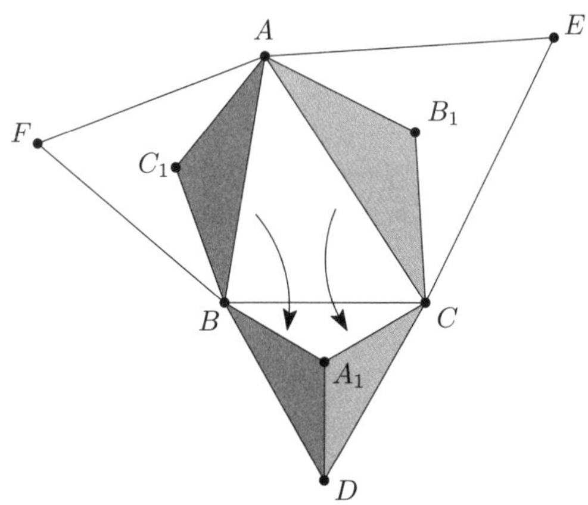

[2]Napoleon Bonaparte (1769–1821) was a French amateur mathematician who sadly chose to win his fame in much less peaceful manner.

Second Proof. We choose to write the similarity of the equilateral triangles in the following order of vertices: $\triangle ABF \sim \triangle ECA \sim \triangle CDB$. By the Averaging Principle, the centroids of the triplets of corresponding points form an equilateral triangle. But those triplets are exactly triangles AEC, BCD, and FAB, so we are done!

37. Let X be a point in the plane of triangle ABC such that

$$\frac{1}{XA} : \frac{1}{XB} : \frac{1}{XC} = a : b : c.$$

Prove that the images of points A, B, C in inversion about X form an equilateral triangle.

Proof. Let r be the radius of inversion and let A', B', C' be the images of points A, B, C, respectively.

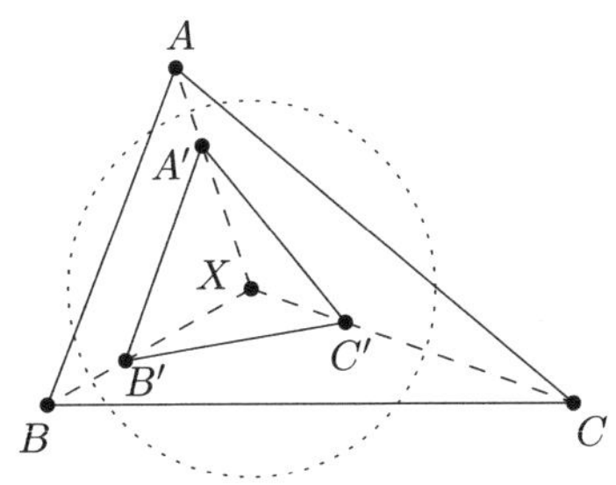

We recalculate distances by Proposition 1.51(b):

$$A'B' = AB \cdot \frac{r^2}{XA \cdot XB}, \quad A'C' = AC \cdot \frac{r^2}{XA \cdot XC}.$$

Comparing shows, we need to prove

$$\frac{AB}{XB} = \frac{AC}{XC},$$

which is just another form of the given

$$\frac{1}{XB} : \frac{1}{XC} = b : c.$$

The equality $A'C' = B'C'$ is proved analogously.

Remark. In fact, for every scalene triangle ABC two such points X exist. More on their existence will be hinted at in the remark following Advanced Problem 12.

38. Let $ABCD$ be a trapezoid such that $BC \parallel AD$ and $\angle CBA = 90°$. Let M be a point on AB satisfying $\angle CMD = 90°$. Let AK be an altitude in triangle DAM and BL an altitude in triangle MBC. Prove that the lines AK, BL, and CD are concurrent.

Proof. Let $X_1 = AK \cap CD$ and $X_2 = BL \cap CD$. Observe that $BL \parallel MD$ as they are both perpendicular to MC. Therefore $\triangle CLX_2 \sim \triangle CMD$ (AA) and we may write

$$\frac{CX_2}{CD} = \frac{CL}{CM} \qquad \text{or} \qquad \frac{CX_2}{X_2D} = \frac{CL}{LM}.$$

Similarly, we obtain

$$\frac{DX_1}{X_1C} = \frac{DK}{KM}.$$

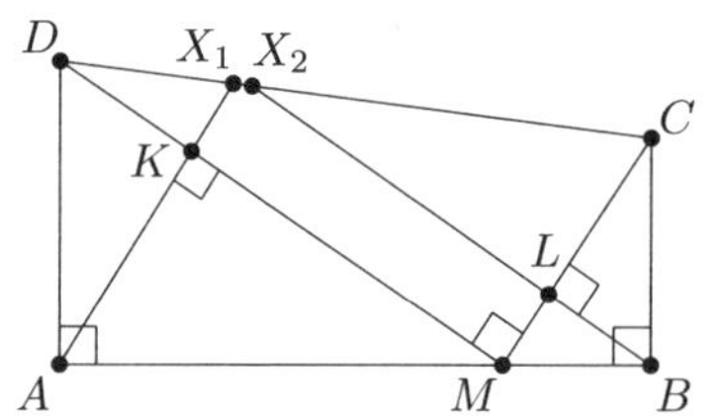

Moreover, $\angle BMC = 180° - 90° - \angle DMA = \angle ADM$, so triangles BMC and ADM are also similar and therefore proportional. Thus

$$\frac{DK}{KM} = \frac{LM}{CL},$$

and this gives us

$$\frac{DX_1}{X_1C} = \frac{DX_2}{X_2C}.$$

Then points X_1, X_2 must coincide, since they divide the segment CD in the same ratio. The conclusion follows.

39. [Poland 2008] An angle with vertex V and a point A in its interior are given. Points X, Y lie on the respective rays of the angle such that $VX = VY$ and the sum $AX + AY$ is the minimal possible. Prove that $\angle XAV = \angle YAV$.

Proof. The question is which pair of points X, Y minimizes the sum $AX + AY$. We learn the answer if we cut away the triangle VXA and glue it on the other side of triangle VAY as triangle VYA'. Now the

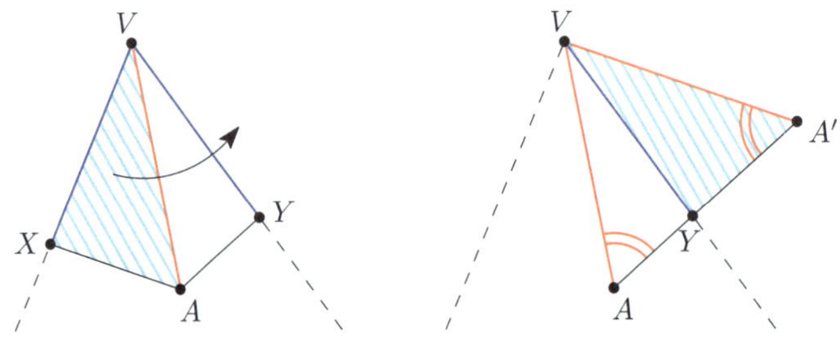

sum $AX + AY$ translates into $AY + YA'$ which, since A and A' are fixed, is minimal when Y lies on AA'. Then as $VA = VA'$, we have $\angle VAY = \angle VA'Y$, which is the same as the desired $\angle XAV = \angle YAV$.

40. [Tournament of Towns 2003] Let ABC be a triangle with $AB = AC$. Let K, L be the points on the sides AB, AC, respectively, such that $KL = BK + CL$. Let M be the midpoint of KL. The line through M parallel to AC intersects BC at N. Find the magnitude of the angle KNL.

Solution. We place line BC horizontally and take a look at horizontal levels of points K, L, and M. Since M is the midpoint, its level is the average of levels of K and L. Moreover, as segments BK, NM, and CL subtend the same angle with BC, their lengths are proportional to their horizontal levels. Hence $MN = \frac{1}{2}(BK + CL)$.

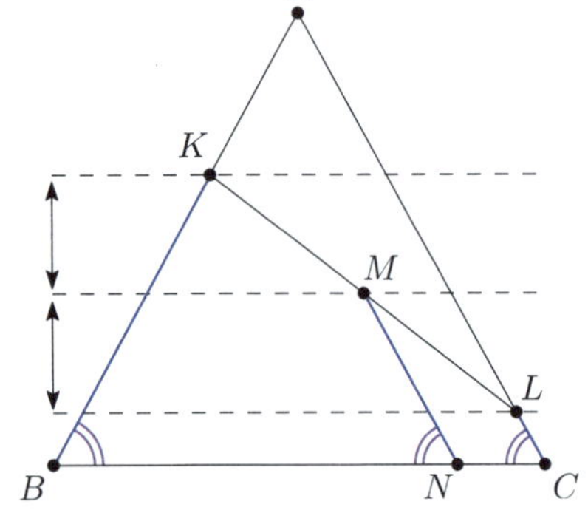

Then the given condition yields $MN = \frac{1}{2}KL = MK = ML$, thus M is the circumcenter of triangle KNL, and as M lies on KL, the angle KNL is right.

41. [based on AIME 2009] Let ABC be a triangle and D the point of contact of the incircle ω with BC. Let DX be a diameter of ω. Show that if $\angle BXC = 90^\circ$, then $5a = 3(b+c)$.

Proof. In triangle BXC with altitude XD we recognize the configuration from Introductory Problem 2, which yields

$$BD \cdot DC = DX^2.$$

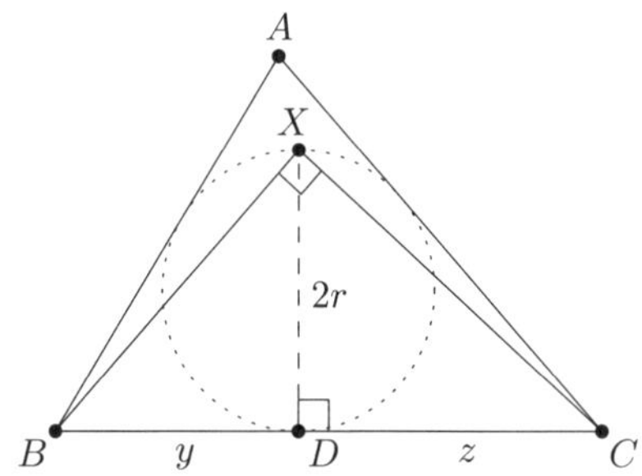

If we recall the xyz formula for inradius r (see Proposition 1.8), the latter turns into

$$yz = 4r^2 = \frac{4xyz}{x+y+z}.$$

After simplification we obtain $y + z = 3x$.

It remains to note that the desired condition $5a = 3(b+c)$ also rewrites as $y + z = 3x$. We are done.

42. [APMO 1994] Given a triangle ABC with circumcenter O, orthocenter H, and circumradius R, prove that $OH < 3R$.

Proof. If ABC is equilateral, then $O = H$ and the conclusion is immediate. Otherwise, the trick is to look at the Euler line of triangle ABC (see Example 1.3).

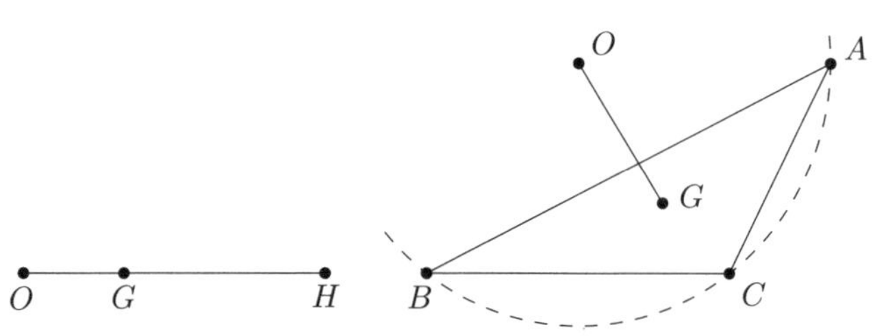

Since the centroid G is in one third from O to H, it suffices to prove $OG < R$. But this is obvious since, the centroid lies always inside the triangle and thus also inside the circumcircle!

43. Circles ω_a, ω_b are internally tangent to a circle ω at distinct points A, B, respectively. Moreover, they are tangent to each other at T. Denote by P the second intersection of AT and ω. Show that BP is perpendicular to BT.

Proof. We draw the common tangent of ω_a and ω_b through T and place it horizontally (with ω_a "above" it).

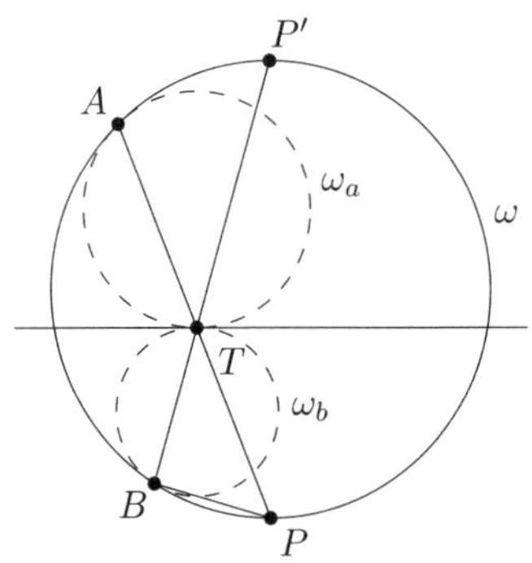

Now consider homothety with center A which takes ω_a to ω. Then T maps to P and since T was the "bottom" point on ω_a, P is the "bottom" point on ω.

Next, we intersect BT with ω for the second time at P' and we may use an analogous argument to show that P' is the "top" point on ω. Then points P and P' form a diameter and $\angle PBT = 90°$.

44. Let ABC be an acute-angled triangle with orthocenter H. Let A', B', C' be the images of A, B, C, respectively, under inversion about H. Prove that H is the incenter of triangle $A'B'C'$. What happens if triangle ABC is obtuse?

Proof. Consider points A, B, and C as pairwise intersections of the circumcircles of triangles BCH, CAH, and ABH. Recall that these circumcircles have equal radii (see Proposition 1.35(d)).

Thus in inversion these circles turn into lines $A'B'$, $B'C'$, and $C'A'$ (see Proposition 1.53) equidistant from H.

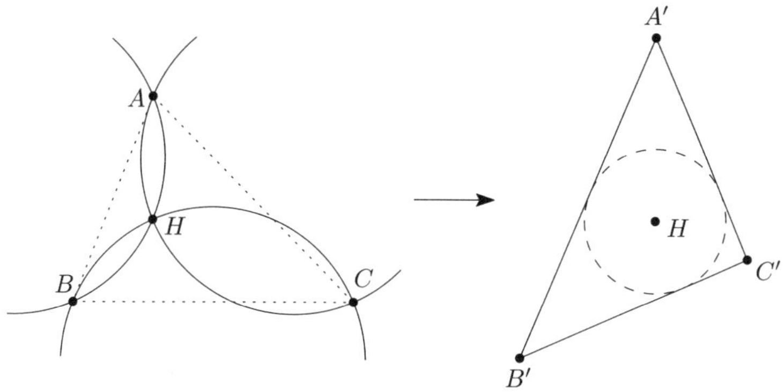

Since triangle ABC was acute, H lies inside triangle $A'B'C'$ and therefore coincides with its incenter. In case of obtuse triangle ABC, H will be one of the excenters in triangle $A'B'C'$.

45. Circles ω_a, ω_b are internally tangent to a circle ω at distinct points A, B, respectively. Moreover, they are tangent to each other at T. Denote by P any intersection of ω and their common tangent through T. Let the lines PA, PB intersect ω_a, ω_b for the second time at X, Y, respectively. Show that XY is a common tangent of ω_a and ω_b.

First Proof. We may assume P is the "top" point of ω in order to ensure an easier visualizing of future homothety arguments.

We observe that since P lies on the radical axis of ω_a and ω_b, the Radical Lemma ensures that $ABYX$ is cyclic (see Proposition 1.23). Now we focus on antiparallel lines in angle APB and we introduce the tangent ℓ to ω at P. Since both ℓ and XY are antiparallel to AB, and ℓ is horizontal, XY is horizontal too.

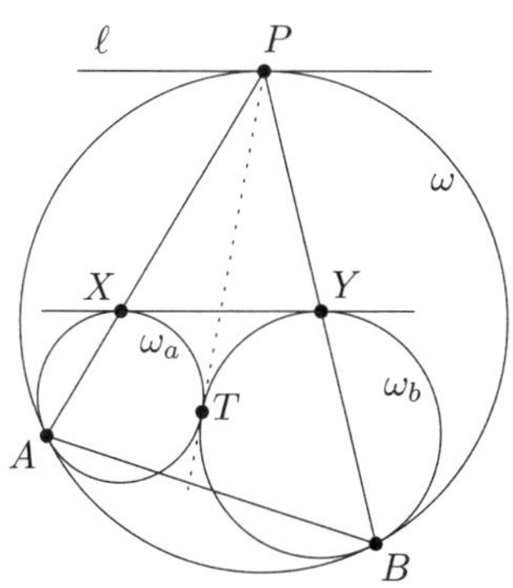

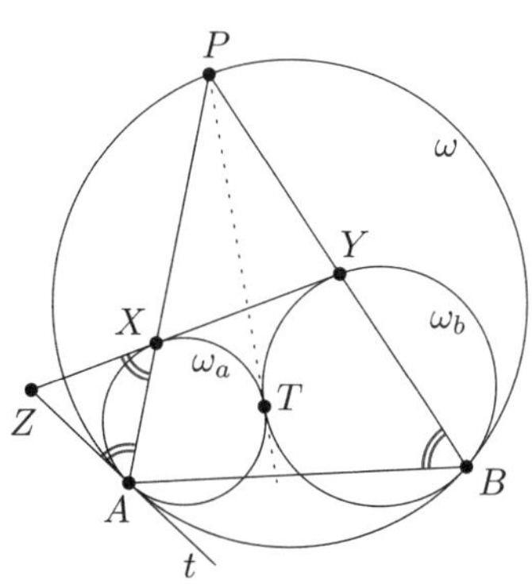

Now the homothety with center A which takes ω to ω_a also takes P to X, so X is the "top" point of ω_a. But then XY is a horizontal line through the "top" point, i.e. a tangent to ω_a at X.

For the same reason, XY is tangent to ω_b at Y.

Second Proof. After we observe $ABYX$ is cyclic as in the first proof, we may also draw the common tangent t to ω and ω_a at A in order to exploit their tangency. For purposes of notation let $Z = t \cap XY$. Then

$$\angle PBA = \angle PAZ \quad \text{and} \quad \angle PBA = \angle ZXA.$$

The equality $\angle PAZ = \angle ZXA$ implies that ZX is tangent ω_a. Similarly, we can show XY is tangent to ω_b.

46. [APMO 1998] Let ABC be a triangle and D the foot of the altitude from A. Let E and F lie on a line passing through D such that AE is perpendicular to BE, AF is perpendicular to CF, and E and F are different from D. Let M and N be the midpoints of the segments BC and EF, respectively. Prove that AN is perpendicular to NM.

Proof. First, we realize this problem is about two circles with diameters AB (call it ω_1) and AC (call it ω_2) intersecting at A and D. This configuration calls for spiral similarity, since the collinearity of E, D, F, and of B, D, C implies (see Proposition 1.48), that spiral similarity centered at A which takes ω_1 to ω_2 takes also triangle AEB to triangle AFC.

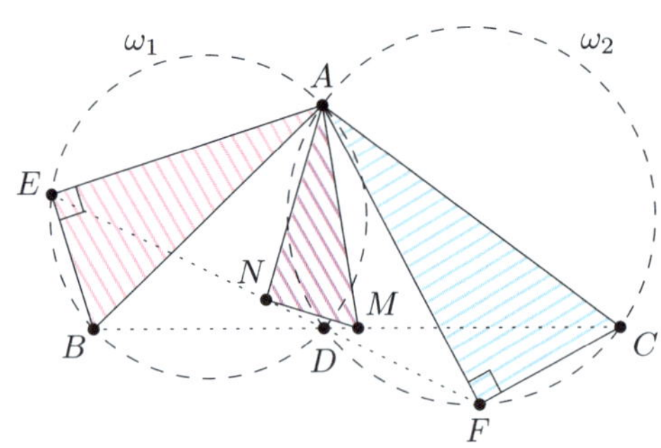

Since the average of these similar triangles is triangle ANM, it has the same shape (recall the Averaging Principle), thus indeed $AN \perp MN$.

47. Four distinct points P, Q, R, and S are given in plane, such that $PQRS$ is not a parallelogram. Find the locus of centers O of rectangles whose sidelines AB, BC, CD, and DA pass through P, Q, R, and S, respectively.

Proof. Denote the midpoints of PR and QS by M, N, respectively (since $PQRS$ is not parallelogram, $M \neq N$).

First let us suppose we found such rectangle $ABCD$. Note that both O and M lie midway between its parallel sides AB and CD, and both O and N lie midway between the sides BC and AD. Thus either $\angle MON = 90°$, or O coincides with one of M and N.

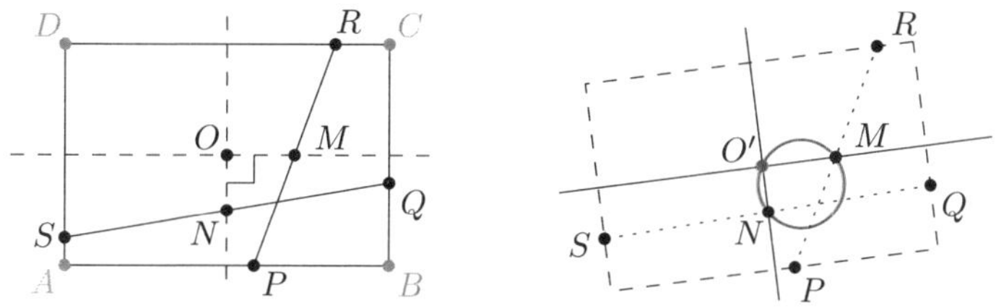

On the other hand, given any point O' on the circle with diameter MN (call it ω), there exists a rectangle $ABCD$ whose sidelines pass through P, Q, R, S, respectively, and whose center is O'. Indeed, lines through P and R parallel to $O'M$, and lines through Q and S parallel to $O'N$ form a rectangle whose midlines are precisely $O'M$ and $O'N$ (if say $O' = N$ we consider line tangent to ω at N instead of $O'N$).

The locus is the circle with diameter MN.

48. Let ω be a circle, BC its fixed chord, and A a variable point on its major arc BC. Let M be the point on the segment AB such that $AM = 2MB$ and let K be the projection of M onto AC. Show that point K moves along a circular arc.

Proof. Since $\angle MAK = \angle BAC$ is fixed as A varies along the arc BC of ω, all the right triangles AKM have the same shape. Even more, since $AM/MB = 2$ is fixed, the shape of all the triangles AKB is the same too. Hence both the ratio BK/BA and the magnitude of the angle ABK are constant implying that the locus of K is simply the image of the locus of A in spiral similarity $\mathcal{S}(B, BK/BA, \angle(BA, BK))$, a circular arc.

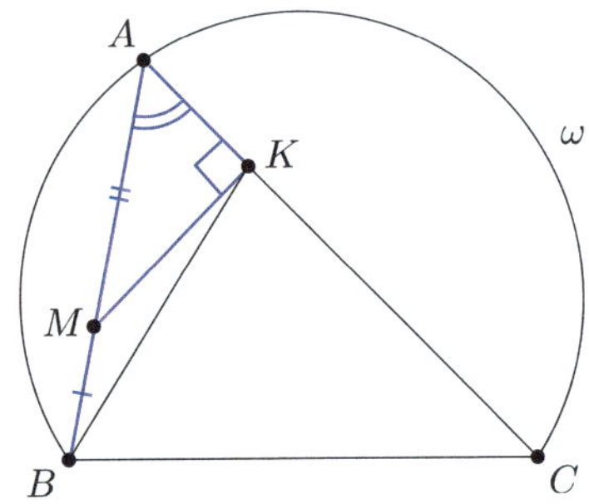

49. In triangle ABC the line isogonal to the median is called the *symmedian*. Let ω be the circumcircle of triangle ABC.

   (a) If $\angle A \neq 90^\circ$ denote by T the intersection of tangents to ω at points B and C. Prove that line AT is the A-symmedian in triangle ABC.

   (b) Let the A-symmedian in triangle ABC meet ω for the second time at S. Prove that
   $$BS \cdot AC = CS \cdot AB.$$

   (c) If X is a point on the symmedian of the triangle ABC and we denote by X_b, X_c its projections on the sidelines AC and AB, then
   $$\frac{XX_b}{XX_c} = \frac{AC}{AB}.$$

First Proof of (a). Assume triangle ABC is acute. We will prove that AT is isogonal with the median in triangle ABC. We draw a line through T, which is antiparallel with BC in $\angle BAC$ and denote its intersections with AB and AC by X and Y, respectively. Our target is to prove that T is the midpoint of XY, since this would ensure AT to be median in triangle AXY and thus also symmedian in triangle ABC.

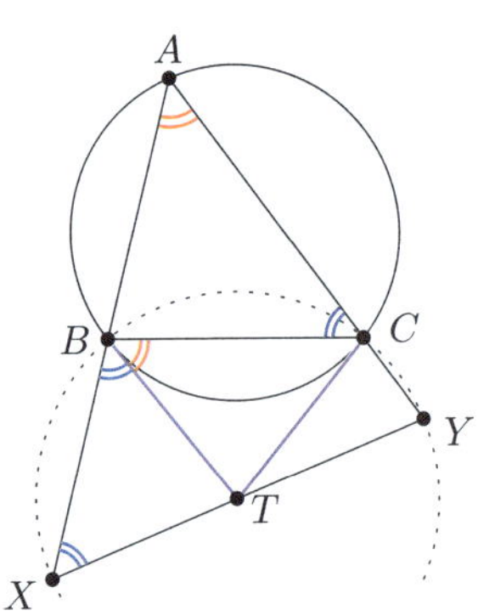

We may as well decide to prove that T is the center of the circumcircle of the cyclic quadrilateral $XYCB$, which has to be the case as $TB = TC$ and we need $TX = TY$. So in fact, the desired conclusion is equivalent to $TX = TB$, which we will show by angle-chasing.

As BC and XY are antiparallel, we have $\angle TXB = \angle C$, and using tangency yields

$$\angle XBT = 180^\circ - \angle A - \angle B = \angle C.$$

Thus $TX = TB$ and the conclusion follows.

In the other cases when ABC is not acute, the proof is analogous.

Second Proof of (a). Perform a $\sqrt{bc}$-inversion. The circle ω will go to the line BC. The lines tangent to ω at B and C will go to circles ω_1 and ω_2 passing through A and tangent to BC at C and B, respectively, and T will go to the second intersection point T' of these circles. Let M be the midpoint of BC. Then the symmedian line AD will go to the median line AM.

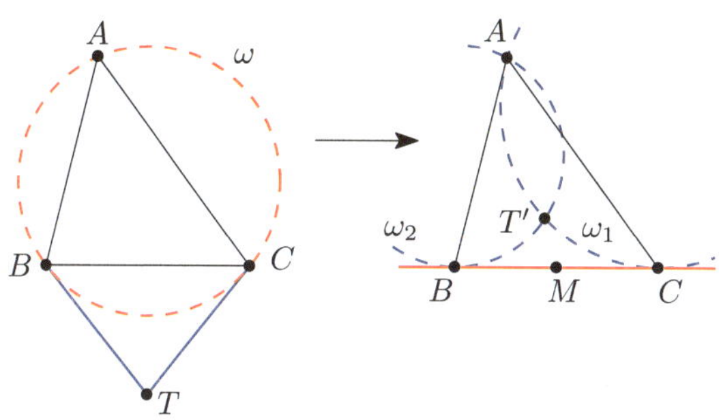

Thus part (a) is equivalent to showing that A, T', and M are collinear. But this is easy since the powers of M with respect to ω_1 and ω_2 are both $MC^2 = MB^2$. Hence M lies on the radical axis AT' of ω_1 and ω_2.

Proof of (b). Let M be the midpoint of BC and R the radius of ω.

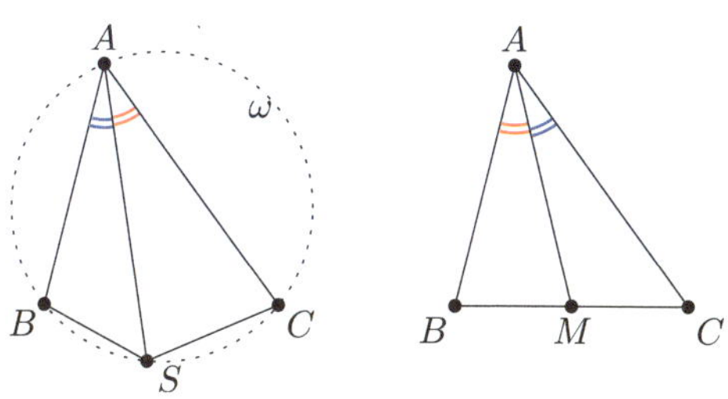

From the Extended Law of Sines we have

$$BS = 2R \sin \angle BAS = 2R \sin \angle CAM$$

and likewise $CS = 2R \sin \angle CAS = 2R \sin \angle BAM$. Hence it suffices to show that $b \sin \angle CAM = c \sin \angle BAM$. However, this follows from the Law of Sines in triangles AMB and AMC, since

$$b \sin \angle CAM = MC \sin \angle AMC = MB \sin \angle AMB = c \sin \angle BAM.$$

Proof of (c). Let's fix the point X and its projections. From the right triangles XX_bA and XX_cA we obtain

$$\frac{XX_b}{XX_c} = \frac{AX \sin \angle XAC}{AX \sin \angle BAX} = \frac{\sin \angle BAM}{\sin \angle CAM}.$$

The problem now reduces to showing $AC \sin \angle CAM = AB \sin \angle BAM$ which we already proved in part (b).

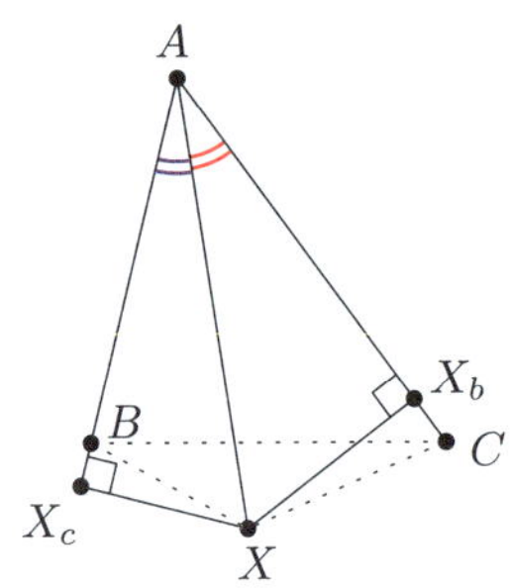

Remark. Also the converse statement is true. If a point X satisfies the equality of ratios, it lies on the symmedian. Can you argue yourself that there cannot two valid points X lying on two different lines through point A?

50. Let A, B, C, and D be distinct points in the plane not lying on one circle. Each set of three points is inverted with respect to the fourth point. Show that the resulting four triangles are mutually similar.

Proof. Realizing it is virtually impossible to draw a reasonable diagram, we decide to make use of the fact that we can calculate the length of every segment after performing inversion (see Proposition 1.51(b)). Indeed, if we denote by B', C', D' the images of B, C, D in inversion with center A and radius 1, we learn that

$$B'C' = \frac{BC}{AB \cdot AC}, \; C'D' = \frac{CD}{AC \cdot AD}, \; D'B' = \frac{BD}{AB \cdot AD}$$

which after taking common denominator of $AB{\cdot}AC{\cdot}AD$ can be rewritten as

$$B'C' : C'D' : D'B' = (AD \cdot BC) : (CD \cdot AB) : (BD \cdot AC).$$

This defines the shape of triangle $B'C'D'$ and since the right-hand side is symmetric in A, B, C, and D, we find that the remaining three triangles also have this shape. We are done.

51. Quadrilateral with escribed circle.

Circle ω is inscribed in angle EAF and is tangent to AE at E and to AF at F. On the segments AE and AF choose points B and D, respectively. Let the tangents from B and D to ω (distinct from AE and AF) intersect at C. Show that:

(a) $AB + BC = CD + DA$.

(b) The incircles of triangles ABD and BCD touch BD at symmetric points with respect to the midpoint of BD.

Proof. In (a), Denote by T, U the points of contact of the lines BC, DC with the circle ω.

By Equal Tangents for vertices B, A and D we find

$$AB + BT = AB + BE = AE = AF = AD + DF = AD + DU.$$

Subtracting $CT = CU$ yields the result.

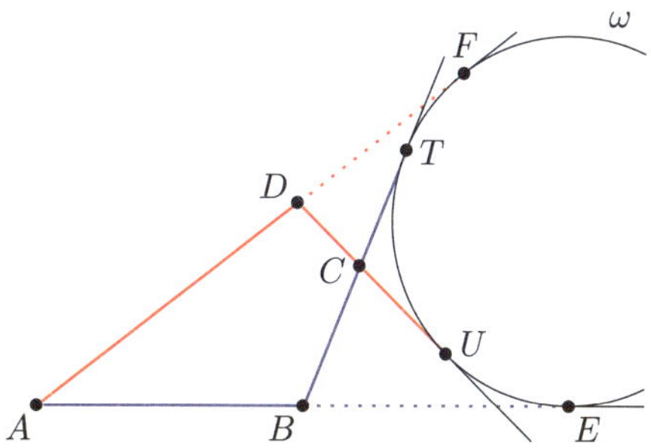

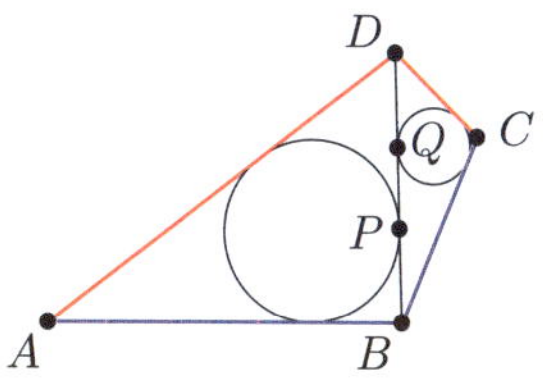

In (b), we work in a figure without circle ω. Denote by P, Q the points of contact of BD with the incircles of triangles ABD, BCD.

By Proposition 1.7 we have

$$BP = \frac{1}{2}(BD + AB - DA) \quad \text{and} \quad DQ = \frac{1}{2}(BD + CD - BC).$$

Since part (a) also reads as $AB - DA = CD - BC$, we obtain $BP = DQ$ implying that P and Q are symmetric about the midpoint of BD.

52. [Euler's Theorem] Triangle ABC is inscribed in circle ω with radius R centered at O. Let I be the incenter of triangle ABC and r its inradius. Prove that $OI^2 = R^2 - 2Rr$.

Proof. We use power of I with respect to the circumcircle. From the very definition we know that $p(I, \omega) = OI^2 - R^2$, hence it suffices to prove that $p(I, \omega) = -2Rr$.

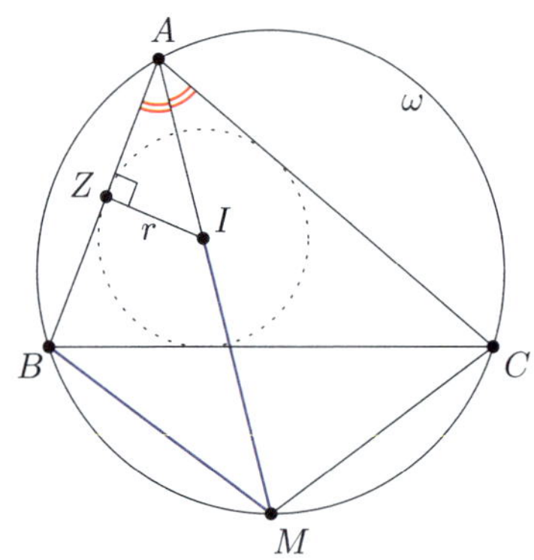

Let M be the second intersection of AI and ω, i.e. the midpoint of arc BC of ω which does not contain A. Since I lies inside ω, we aim to prove $IA \cdot IM = 2Rr$.

We know that $MI = MB$ (see Proposition 1.38(b)) and thus by the Extended Law of Sines in triangle AMB, we have $MI = MB = 2R\sin\frac{\angle A}{2}$.

As for the distance IA, we introduce the point of contact Z of the incircle with AB and use right triangle AIZ from which we obtain

$$AI = \frac{r}{\sin\frac{\angle A}{2}}.$$

Together we obtain $MI \cdot AI = 2rR$ and we may conclude.

Remark. Very analogous argument can be applied to show that $OI_a^2 = R^2 + 2Rr_a$, where I_a is the center of A-excircle of triangle ABC and r_a its radius. We encourage the reader to verify this.

53. Customizing inversion.

(a) Let ω be a circle and I a point in its exterior. Prove that there exists a circle i centered at I such that ω is preserved in inversion about i.

(b) Let ω_1, ω_2, ω_3 be three circles with non-collinear centers, each outside of the other. Prove that there exists a circle i such that inversion about i preserves ω_1, ω_2, and ω_3.

First Proof of (a). Denote the tangents to ω passing through I by ℓ_1, ℓ_2 and the respective points of tangency by T_1, T_2. Since any inversion about I preserves ℓ_1 and ℓ_2, it maps ω to a circle inscribed in the angle formed by ℓ_1 and ℓ_2. By letting the radius of i be equal to $IT_1 = IT_2$ we ensure that T_1 and T_2 are preserved and since there is unique circle tangent to ℓ_1 at T_1 and to ℓ_2 at T_2, the circle ω is preserved too.

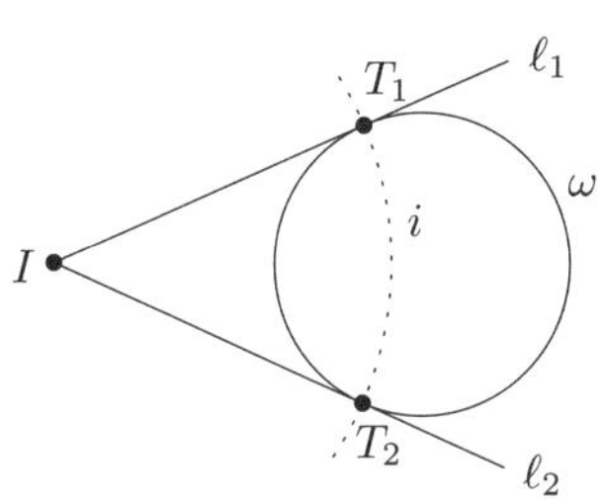

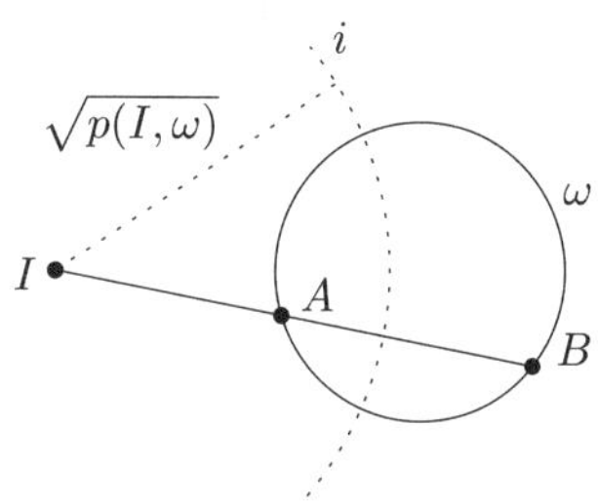

Second Proof of (a). We offer another approach which we will follow in the next part too. Let ℓ be any line passing through I and intersecting ω at (not necessarily distinct) points A, B. As the product $IA \cdot IB = p(I,\omega)$ is constant, it suffices to let the radius of i be equal to $r_i = \sqrt{p(I,\omega)}$ since then we have

$$IA' = \frac{r^2}{IA} = \frac{IA \cdot IB}{IA} = IB$$

which implies that A is mapped to B and vice versa.

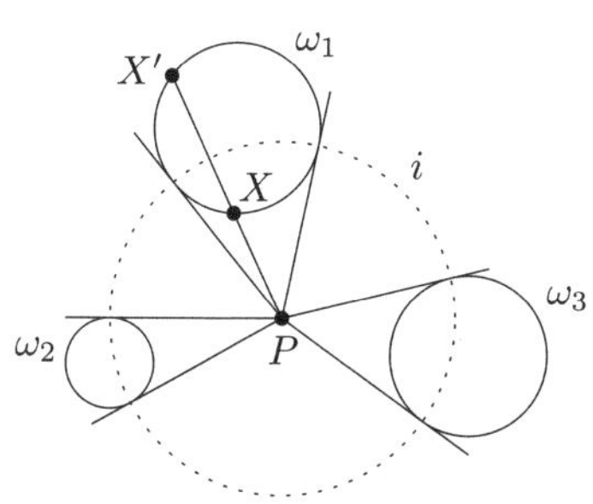

Proof of (b). Let P be the radical center (see Proposition 1.22) of ω_1, ω_2, and ω_3. Since the circles lie outside each other, point P lies outside them too and $p(P, \omega_1) = p(P, \omega_2) = p(P, \omega_3) = p > 0$. As in the second proof of part (a) we conclude that the circle with center P and radius $\sqrt{p}$ has the desired property.

Chapter 5

Solutions to Advanced Problems

1. In acute triangle ABC let E, F be the points of contact of the incircle with the sides AB, AC, respectively, and let L and M be the feet of B and C-altitudes. Show that the incenter I' of triangle ALM coincides with the orthocenter H' of triangle AEF.

   Proof. We draw two separate diagrams and prove that I' and H' lie on the same ray from A and with the same distance from it.

   First, we focus on I'. This certainly lies on the bisector of $\angle A$ and recalling that the factor of similarity between triangles ABC and ALM is $|\cos \angle A|$ (see Proposition 1.35(e)), we can write $AI' = AI \cdot |\cos \angle A|$, where I is the incenter of triangle ABC.

   For H', we first note that triangle AEF is isosceles, thus its altitude is also the bisector of $\angle A$. The distance AH' can be found as $AH' = 2R|\cos \angle A|$ (see Proposition 1.35(f)), where $2R$ is the circumdiameter of triangle AEF. But as points E and F lie on a circle with diameter AI, this circumdiameter is actually AI and the conclusion follows.

2. In triangle ABC with $\angle BAC = 120°$, denote by D, E, F the intersections of the respective angle bisectors with the opposite sides BC, CA, AB. Find $\angle EDF$.

   Solution. Focus on triangle ADC. Observe that AF is its external angle bisector by A and CF is its internal angle bisector. Thus, F is inevitably the C-excenter of triangle ACD. Likewise, E is the B-excenter of triangle ABD. Lines DF and DE then bisect the angles ADB and CDA, making angle DEF one half of a straight angle, i.e. $90°$.

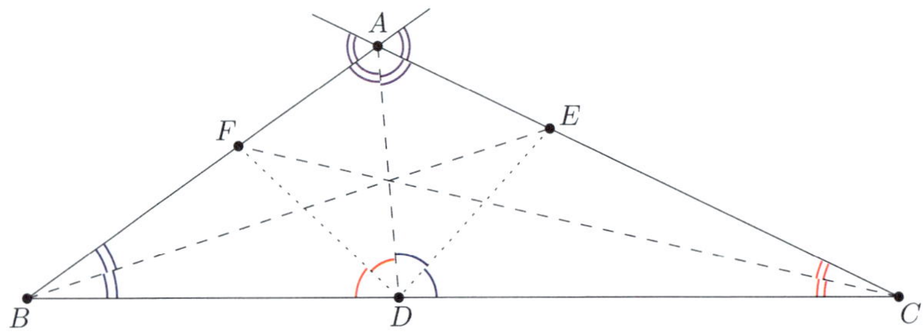

3. [Romania 2006] Let ABC be a triangle with $AB = AC$. Let D be the midpoint of BC, M the midpoint of AD and N the projection of D onto BM. Prove that $\angle ANC = 90°$.

   First Proof. Draw point X such that $ADCX$ is a rectangle. Then B, M, and X are collinear – indeed, $\triangle BDM \sim \triangle BCX$ (SAS), thus the triangles are homothetic from B. As $\angle DNX = 90° = \angle DAX$, it follows that N lies on the circumcircle of the rectangle $ADCX$. Since the other diagonalAC is also a diameter of that circle, we have $\angle ANC = 90°$.

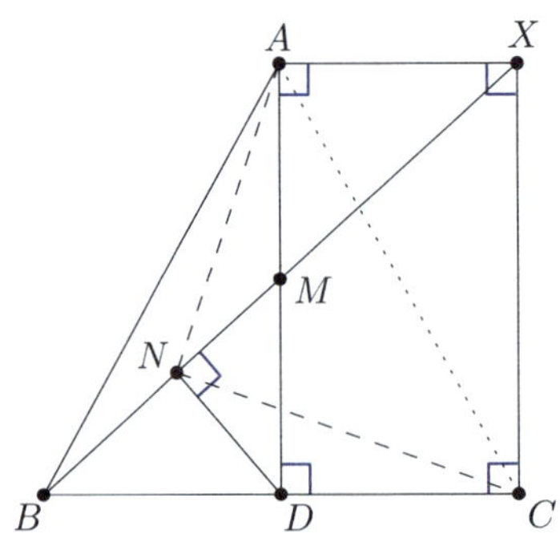

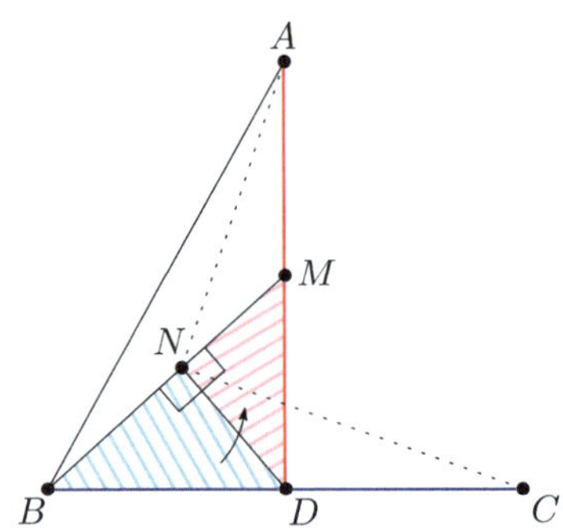

   Second Proof. Realizing that $\triangle BND \sim \triangle DNM$ (AA), we see that the spiral similarity $\mathcal{S}(N, ND/NB, +90°)$ takes the segment BD to the segment DM. Since the triplets of points B, D, C, and D, M, A have the same shape, $\mathcal{S}$ also sends C to A and $\angle ANC = 90°$ follows.

4. Let ABC be an acute-angled triangle with $\angle A = 60°$ and $AB > AC$. Let I be its incenter.

   (a) If H is the orthocenter of triangle ABC, prove that

   $$2\angle AHI = 3\angle B.$$

   (b) If M the midpoint of AI, prove that M lies on the nine-point circle[1] of triangle ABC.

Proof.

(a) (APMO 2007) The angle $\angle AHI$ is not directly accessible so it is natural to expect some circle to arise.

We recall basic angles $\angle BIC = 90° + \frac{1}{2}\angle A = 120°$ (see Proposition 1.38) and $\angle BHC = 180° - \angle A = 120°$ (see Proposition 1.35(c) for acute-angled triangles).

Thus $BCHI$ is cyclic (with vertices in this order due to $AB > AC$) and we may now finish the problem easily. Indeed, using the angle by H one more time we learn

$$\angle IHC + \angle CHA = \left(180° - \frac{1}{2}\angle B\right) + (180° - \angle B)$$

and hence $\angle AHI = \frac{3}{2}\angle B$.

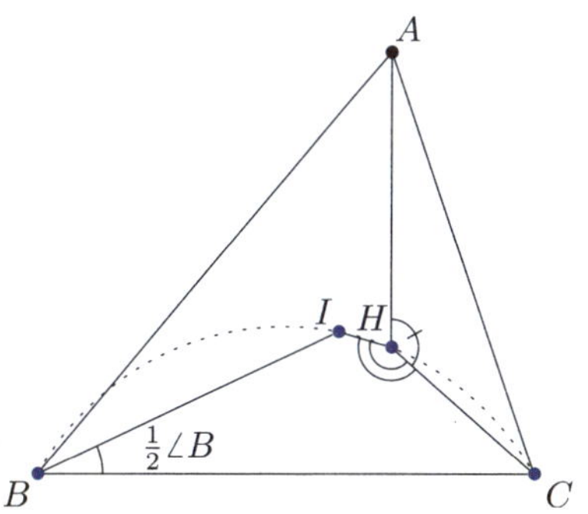

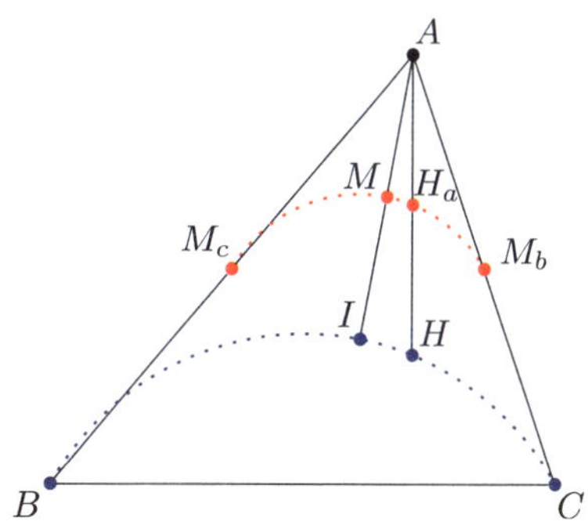

[1]For explanation see Theorem 1.37.

(b) Once we observed that $BCHI$ is cyclic, we just need to realize that homothety $\mathcal{H}(A, \frac{1}{2})$ takes triangle BHC to triangle $M_cH_aM_b$, where M_c, H_a, and M_b are the midpoints of AB, AH, and AC, respectively. Thus the circle BHC is taken to the nine-point circle of triangle ABC (see Theorem 1.37) and since I is taken to M, we may conclude.

5. [Brazil 2008] Quadrilateral $ABCD$ inscribed in a circle ω contains its center O in its interior. Let r and s be the lines obtained by reflecting AB with respect to the internal bisectors of $\angle CAD$ and $\angle CBD$, respectively. If P is the intersection of r and s, prove that OP is perpendicular to CD.

Proof. Note that bisectors of both angles $\angle CAD$ and $\angle CBD$ intersect the circle at the same point, namely at the midpoint E of arc CD (not containing A and B). Now, since $OE \perp CD$, we may erase points C and D and aim to prove that O, E, and P are collinear.

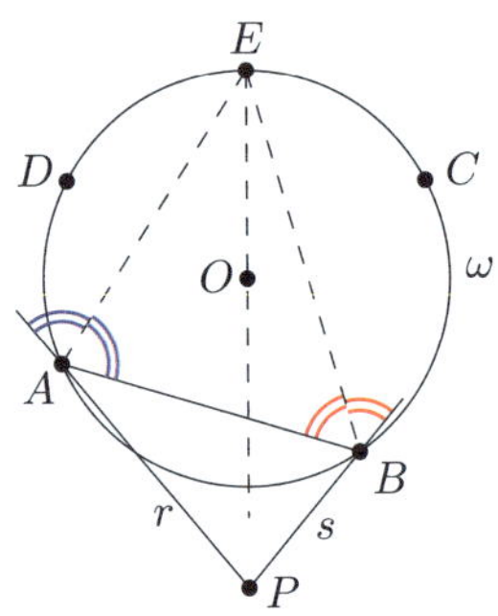

As O lies inside $ABCD$, angle AEB is acute and thus $\angle BAE + \angle ABE > 90^\circ$ implying that lines AE and BE bisect external (and not internal) angles in triangle APB. Therefore E is the P-excenter in this triangle.

We recognize ω as part of the Big Picture from Proposition 1.42(b) for triangle APB and recall that its center lies on the angle bisector of $\angle BPA$. The conclusion follows as E, O, P form the angle bisector of $\angle BPA$.

6. Let X be the foot of perpendicular from vertex B of the triangle ABC ($AB < AC$) to the angle bisector of $\angle A$.

   (a) Let M, P be the midpoints of AB, BC, respectively. Prove that X lies on MP.

(b) Let D, E be the points of contact of the incircle with sides BC, AC, respectively. Prove that X lies on the segment DE.

Proof.

(a) To prove that X lies on MP is the same as to prove that it lies half the way from B to line AC (consider the homothety $\mathcal{H}(B,2)$). Denote by X' the intersection of BX and AC.

We draw AX vertically and observe that since BX' is horizontal, it cuts off an isosceles triangle from angle BAC. Thus $BX = XX'$ and we are done.

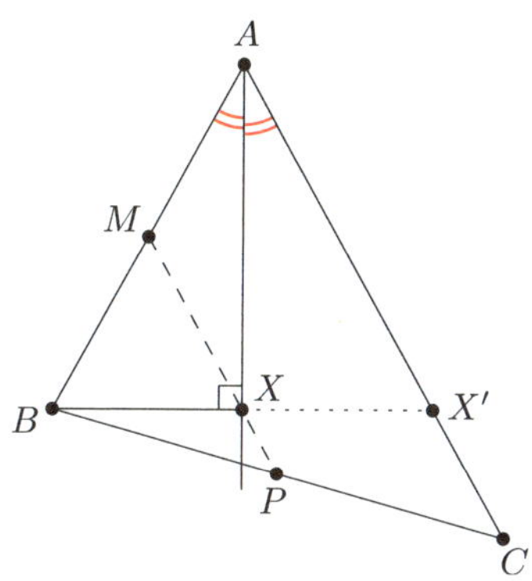

(b) We seek the connection between the points of contact of the incircle and the point X. Let I be the incenter of triangle ABC.

Then $\angle IDB = \angle IXB = 90°$ so the points B, D, X, I are concyclic (thanks to $AB < AC$ in this order of vertices). Now it is straightforward to express $\angle XDB$ and $\angle EDB$ in terms of $\angle A$, $\angle B$, $\angle C$.

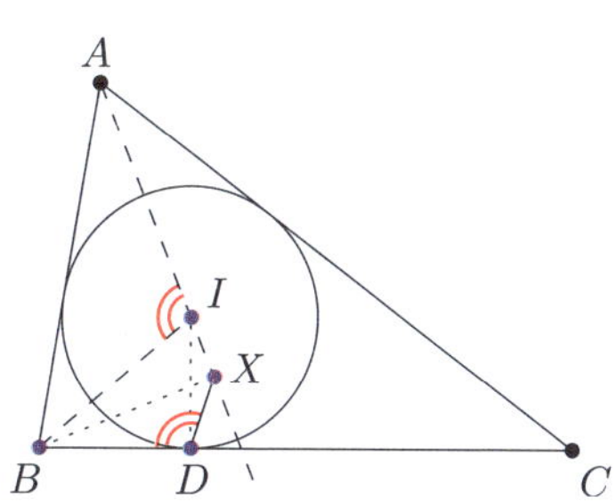

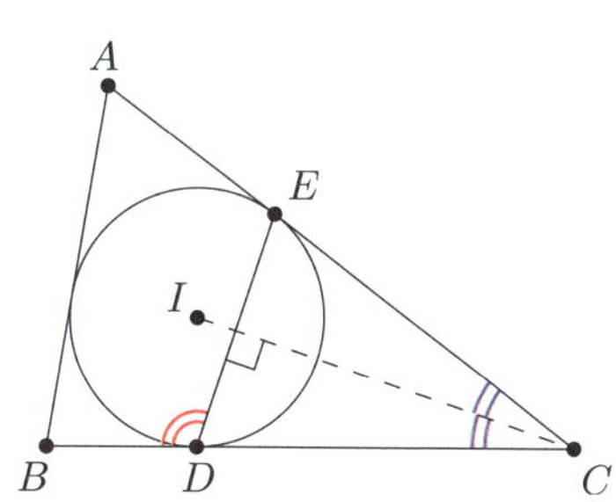

Using the concyclicity we obtain $\angle XDB = \angle AIB = 90° + \frac{1}{2}\angle C$ (recall Proposition 1.38(a)) and $\angle EDB$ can be calculated as an external angle in one half of the isosceles triangle DCE as $90° + \frac{1}{2}\angle C$ Thus, the lines DX and DE coincide and we are done.

7. [Junior Balkan 2010] Let BK and CL be angle bisectors in an acute triangle ABC with incenter I (K lies on the side AC, L lies on the side AB). The perpendicular bisector of LC intersects the line BK at point M. Point N lies on the line CL such that NK is parallel to LM. Prove that $NK = NB$.

   Proof. We identify points M and N as midpoints of arcs in some circumcircles.

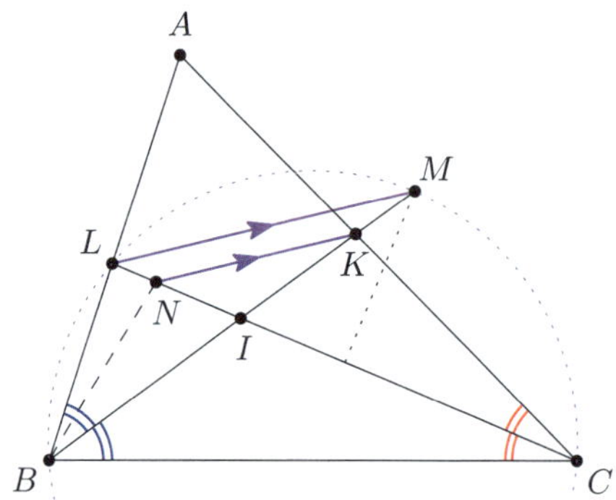

   Namely, since M is the intersection of the bisector of $\angle CBL$ and the perpendicular bisector of LC, it is the midpoint of the minor arc LC of the circumcircle of triangle LBC. In particular, $BCML$ is cyclic.

   But then also $BCKN$ is cyclic, since in $\angle BIC$ the line LM is antiparallel to BC and has the same direction as NK. Finally, as N is the intersection of the circumcircle of triangle BCK and the bisector of $\angle C$ it is the midpoint of minor arc BK, which ensures $NK = NB$.

8. [All-Russian Olympiad 2001] Circles ω_1, ω_2 with radii R_1 and R_2 are internally tangent at N (with ω_1 inside ω_2). Let K be an arbitrary point on ω_1. The tangent to ω_1 at K intersects ω_2 at A and B. Let M be the midpoint of the arc AB of ω_2 not containing point N. Prove that the circumradius R of triangle KBM does not depend on the choice of K.

   Proof. First, place AB horizontally with N "above" it and observe that as K and M are the "bottom" points of the circles ω_1, ω_2, they are collinear with the center N of homothety which sends ω_1 to ω_2 (see Example 1.4 if needed).

   Next, we denote the angle MKB by φ and observe that φ corresponds to some arcs in all three circles. In ω_1 it due to tangency corresponds to arc NK, in ω_2 it corresponds to the sum of arcs BM and AN (see Corollary 1.14(a)), which equals arc NM, and of course in the circumcircle

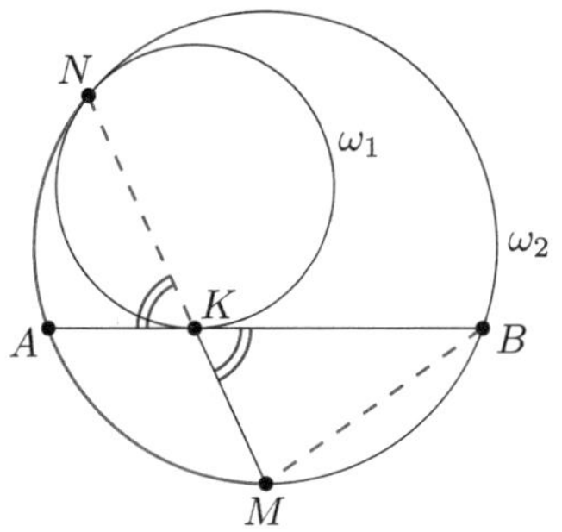

of triangle KBM it corresponds to MB. Using the Extended Law of Sines and the Shooting Lemma (see Proposition 1.40(a)), we may thus calculate R as

$$\begin{aligned}(2R)^2 &= \frac{MB^2}{\sin^2\varphi} = \frac{MK\cdot MN}{\sin^2\varphi}\\ &= \frac{MN}{\sin\varphi}\cdot\left(\frac{MN}{\sin\varphi}-\frac{NK}{\sin\varphi}\right) = 2R_2(2R_2-2R_1) = 4R_2(R_2-R_1),\end{aligned}$$

which is indeed independent of the choice of K.

9. The external common tangent of the circles Γ_1, Γ_2 with centers O_1, O_2 is tangent to them at distinct points A_1, A_2, respectively. The circle with diameter A_1A_2 meets Γ_1, Γ_2 for the second time at B_1, B_2, respectively. Prove that the lines A_1B_2, B_1A_2 and O_1O_2 are concurrent.

Proof. Let A_1A_2 be horizontal with ω_1, ω_2 "above" it. We will guess the common point.

Extend A_1B_2 to meet Γ_2 for the second time at C_2. Since $\angle A_1B_2A_2 = 90^\circ$, we have $\angle A_2B_2C_2 = 90^\circ$ implying that A_2 and C_2 are antipodal points of Γ_2. In other words, C_2 is the "top" point on Γ_2.

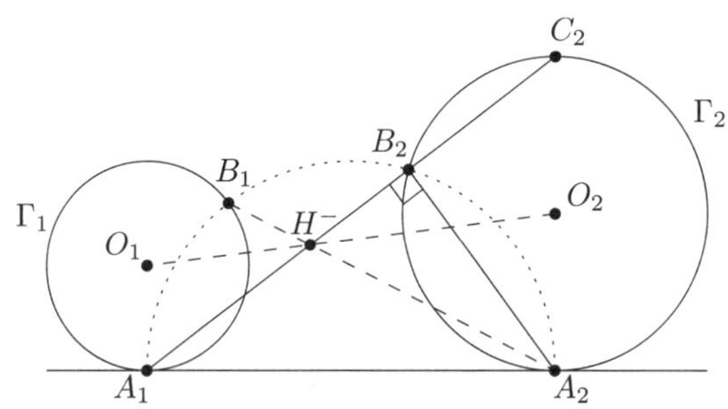

Now the natural choice for the point of concurrence is the center H^- of the negative homothety that maps Γ_1 to Γ_2. As A_1 and C_2 correspond in this homothety, line A_1B_2 passes through H^-. By precisely the same argument, line A_2B_1 passes through H^- too. Finally, H^- clearly lies on O_1O_2, which finishes the proof.

10. [Poland 2000] A circle passing through the vertex A of a parallelogram $ABCD$ intersects the segments AB, AC, AD for the second time at P, Q, R, respectively. Prove that

$$AP \cdot AB + AR \cdot AD = AQ \cdot AC.$$

Proof. The metric relation looks somewhat similar to Ptolemy's Inequality (see Theorem 1.46) in its equality case.

The (real) Ptolemy's Inequality applied to cyclic quadrilateral $APQR$ states

$$AP \cdot QR + AR \cdot PQ = AQ \cdot PR.$$

If $AB/QR = AD/PQ = AC/PR = k$ were true, then the result would follow just by multiplying by k. As $AB = DC$, this is equivalent to $\triangle ADC \sim \triangle PQR$.

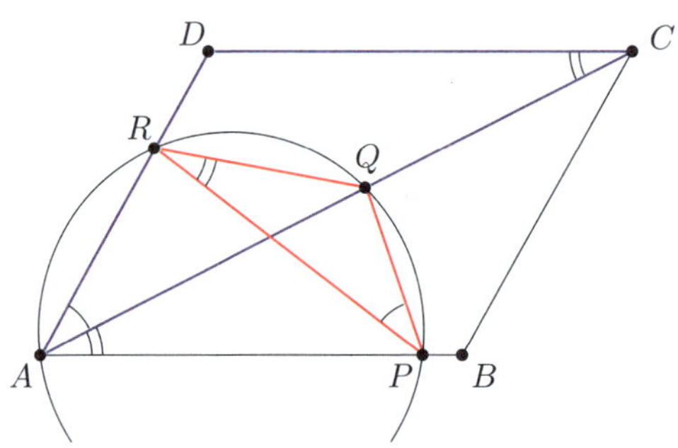

Perhaps surprisingly, this similarity is quickly obtained by AA, since the cyclic quadrilateral $APQR$ gives $\angle QPR = \angle QAR = \angle CAD$ and $\angle PRQ = \angle PAQ = \angle ACD$.

11. A scalene triangle ABC with incenter I and $D = AI \cap BC$ satisfies $b + c = 2a$. Show that:

(a) $GI \parallel BC$, where G is the centroid of triangle ABC.

(b) $\angle OIA = 90°$, where O is the circumcenter of triangle ABC.

(c) Let E and F be the midpoints of AB and AC, respectively. Then I is the circumcenter of triangle DEF.

Proof. Statements of the problem lead us to the belief that ratios on the angle bisector AD have very special values in this kind of triangle. Let's first focus on those ratios. For the sake of simplicity, we may assume $DI = 1$.

As the incenter divides the angle bisector AD in the known ratio (see Proposition 1.38(c)), we find

$$\frac{AI}{ID} = \frac{b+c}{a} = 2.$$

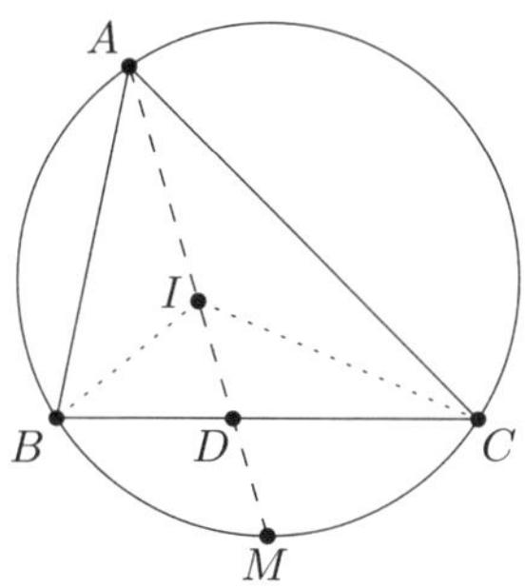

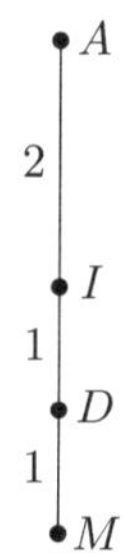

Next, we denote by M the midpoint of arc BC (not containing A) of the circumcircle of triangle ABC. We want to find MI. We recall the Shooting Lemma (see Proposition 1.40(b)), which gives $MI^2 = (MI - 1) \cdot (MI + 2)$ and thus $MI = 2$ and so $MD = 1$.

Now we know enough about ratios and we may proceed to the problem itself.

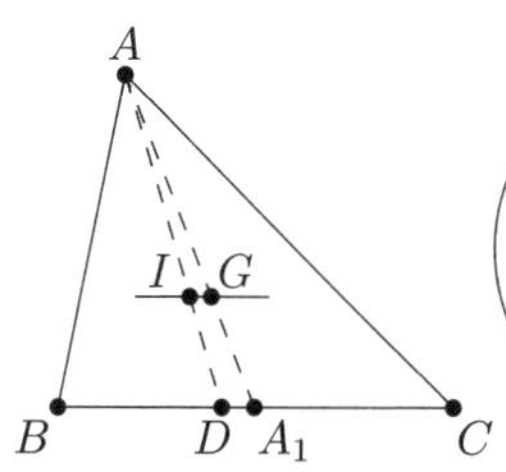

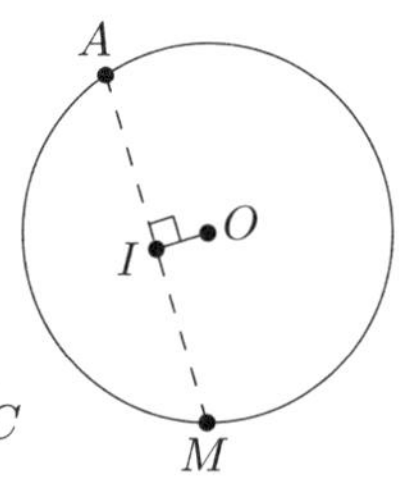

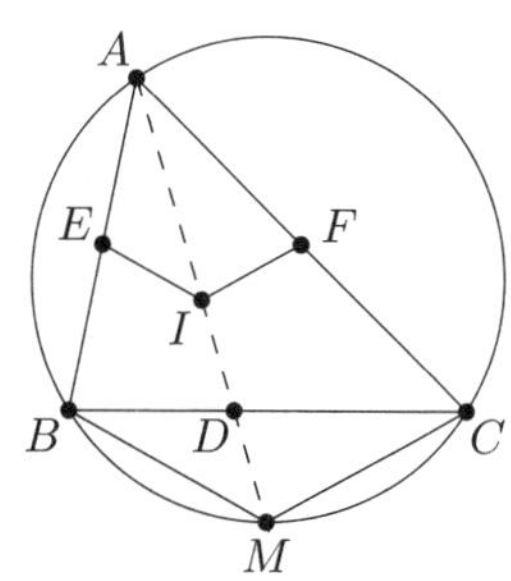

(a) Since I divides AD in ratio $2:1$ and G divides the median AA_1 ($A_1 \in BC$) in ratio $2:1$, the homothety $\mathcal{H}(A, \frac{3}{2})$ takes IG to DA_1 and thus $BC \parallel GI$.

(b) Since I is the midpoint of the chord AM, we indeed have $\angle OIA = 90^\circ$.

(c) We will prove $IE = IF = ID = 1$. As IE is a midline in triangle ABM, we have $IE = \frac{1}{2}MB = \frac{1}{2}MI = 1$ (recall Proposition 1.38(b)). Same argument shows $IF = 1$ and we are done.

12. Points B, D, and C are collinear in this order and $BD < DC$. Find the locus of points X such that $\angle BXD = \angle DXC$.

Solution. Assume we found such point X. Being disappointed that the equal angles intercept distinct segments, we decide to map one segment on the other.

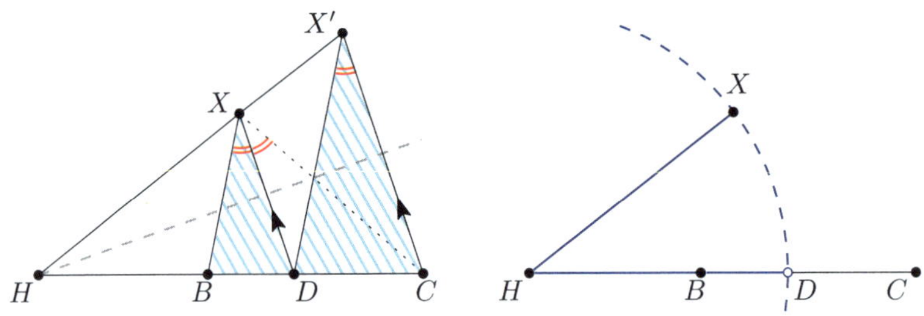

Consider positive homothety $\mathcal{H}$ sending BD to DC and its center $H \in BC$. If X' is the image of X in $\mathcal{H}$, then

$$\angle DX'C = \angle BXD = \angle DXC$$

and as expected $DCX'X$ is cyclic. Moreover, as $DX \parallel CX'$, it is an isosceles trapezoid. Thus, from symmetry of the trapezoid, we have $HD = HX$, which implies that X runs along a circle centered at H with radius HD. By reversing the chain of arguments, we can see that every point $X \notin BC$ of the circle satisfies the desired $\angle BXD = \angle DXC$.

Remark. We have in fact solved a classical problem from triangle geometry: Given triangle ABC, what is the locus of points X for which

$$\frac{XB}{XC} = \frac{AB}{AC}?$$

If D is the foot of the A-angle bisector, then due to the Angle Bisector Theorem this rewrites as

$$\frac{XB}{XC} = \frac{DB}{DC},$$

which happens if and only if $\angle BXD = \angle DXC$ (again by the Angle Bisector Theorem) as in our problem. Thus the answer is the circle we have just found – the so-called *Apollonius*[2] *circle* of triangle ABC with respect to vertex A. Apollonius' circles corresponding to vertices B and C can be defined analogously. We encourage the reader to verify that these three circles intersect at two common points which have the property from Introductory Problem 37.

13. Let ABC be a triangle and P a variable point on the arc AB of its circumcircle ω not containing point C. Let X, Y be the points on the rays BP, CP such that $BX = AB$ and $CY = AC$, respectively. Prove that all such lines XY pass through a fixed point independent of the choice of P.

First Proof. What happens to X when P moves along the arc AB? Since the distance BX is fixed, point X runs along (fixed) circle ω_b with center B passing through A. Likewise, Y traces an arc of a circle ω_c with center C and passing through A. Moreover, since

$$\angle ABX = \angle ABP = \angle ACP = \angle ACY,$$

the triangles ABX and ACY are directly similar (SAS) and the spiral similarity centered at A which maps ω_b to ω_c and B to C maps also X to Y. Hence the line XY passes through the second intersection of ω_b and ω_c, i.e. the reflection A' of A about BC (see Proposition 1.48).

Second Proof. As in the first proof (but without actually drawing the circles) we note that the triangles ABX and ACY are isosceles and similar. Hence it is natural to consider spiral similarity centered at A which maps triangle ABX to triangle ACY.

By fixing point P, we fix the shape of those triangles and observe that as B "glides" to C, point X "glides" to Y. In other words, line XY is the locus of points Z for which there exists point D on the line BC such that triangle AZD is similar to both ABX and ACY. But the reflection A' of A about BC clearly has this property! Hence all the lines XY pass through A'.

[2]Apollonius of Perga (c. 240 – 190 BC) was a great greek geometer known for his work on conic sections.

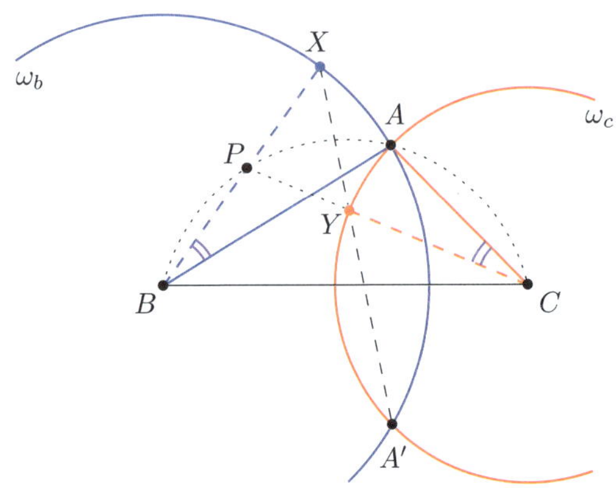

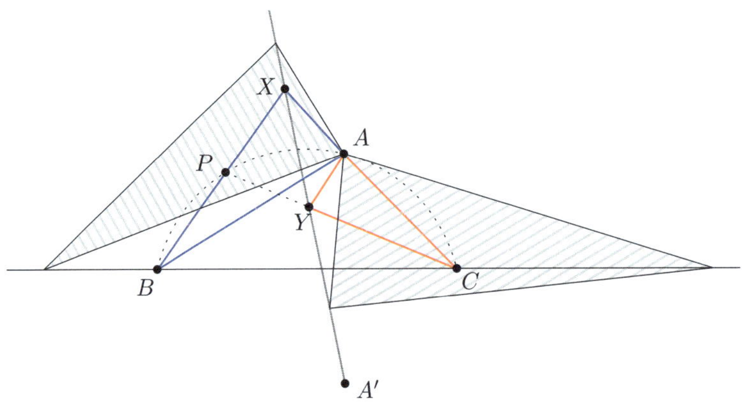

Third Proof. If we are aware of the circles ω_b and ω_c from the first proof and manage to guess the common point would be A', we may also verify it by angle-chasing.

We have

$$\angle XA'A = \frac{1}{2}\angle XBA \equiv \frac{1}{2}\angle PBA = \frac{1}{2}\angle PCA \equiv \frac{1}{2}\angle YCA = \angle YA'A,$$

hence X, Y and A' are collinear and we are done.

14. [AIME 2007] Four circles ω, ω_a, ω_b, ω_c with the same radius are drawn in the interior of triangle ABC such that ω_a is tangent to the sides AB and AC, ω_b to BC and BA, ω_c to CA and CB, and ω is externally tangent to ω_a, ω_b, and ω_c. If the side lengths of triangle ABC are 13, 14, and 15, determine the radius of ω.

Solution. In order to make use of the equal radii we have to introduce some new points. Denote by A', B', C', O the centers of the circles ω_a, ω_b, ω_c, ω, respectively, and by x their common radius.

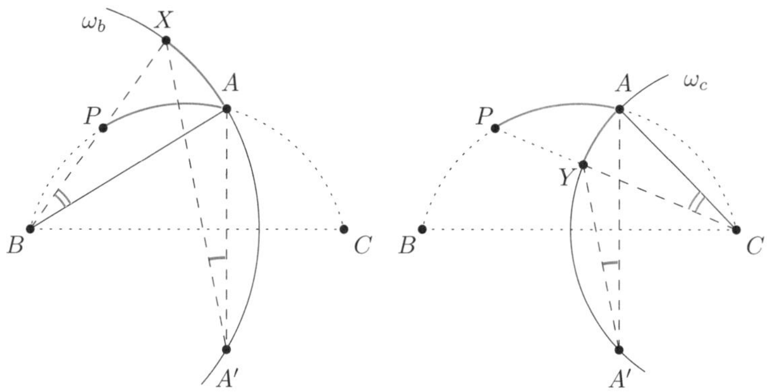

Since the radii of ω_b and ω_c are the same, points B' and C' have the same distance from the line BC and so $B'C' \parallel BC$. The same holds for the other sides, and thus the triangles ABC and $A'B'C'$ are similar.

Recall that the perimeter, area, inradius, circumradius or almost anything in triangle ABC can be calculated given its sides. If we were able to express two such quantities in triangle $A'B'C'$ in terms of x, we would equate their ratios and obtain the answer ("similar means proportional").

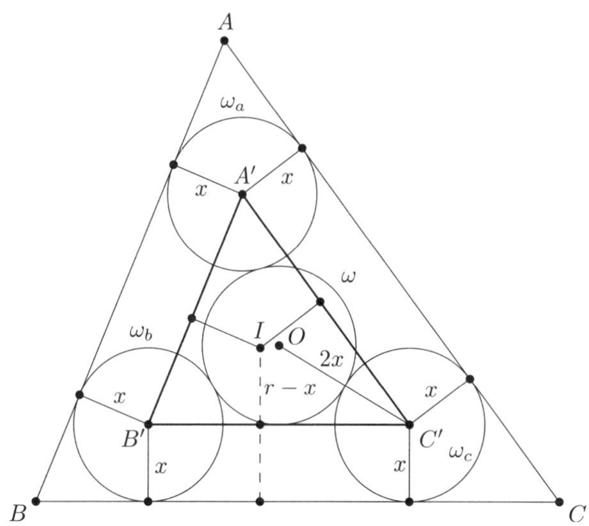

As $OA' = OB' = OC' = 2x$, point O is the circumcenter of triangle $A'B'C'$ and its circumradius equals $2x$.

Moreover, denote by I the incenter of triangle ABC and by r its inradius. The distance of I to all the sides of triangle $A'B'C'$ equals $r - x$, hence I is also the incenter of triangle $A'B'C'$ and its inradius equals $r - x$.

On the other hand, using xyz formulas for triangle ABC (see Proposition

1.8) we compute $r = 4$ and $R = \frac{65}{8}$. Thus it suffices to solve

$$\frac{\frac{65}{8}}{4} = \frac{2x}{4-x},$$

which yields $x = \frac{260}{129}$.

15. Broken circle.

(a) Point P inside a parallelogram $ABCD$ satisfies $\angle BPC + \angle DPA = 180^\circ$. Prove that $\angle CBP = \angle PDC$.

(b) Let $ABCD$ be a trapezoid with $AB \parallel CD$ and $AB > CD$. Points K and L lie on the line segments AB and CD, respectively, such that $\frac{AK}{KB} = \frac{DL}{LC}$. Suppose that there are points P and Q on the line segment KL satisfying $\angle APB = \angle DCB$ and $\angle CQD = \angle CBA$. Prove that the points P, Q, B, and C are concyclic.

Proof.

(a) The constraint reminds us of cyclic quadrilaterals, so we try to create one. We take the triangle APD and translate it so that A goes to B and D to C. Then, if we denote by P' the image of P, the quadrilateral $PBP'C$ is cyclic.

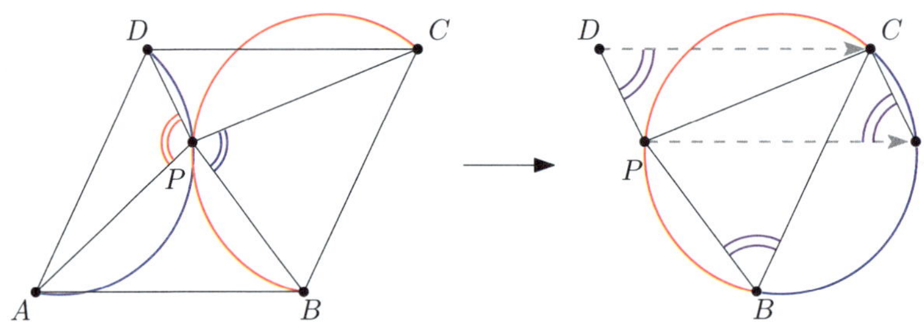

Combining this with the fact that $PP'CD$ is a parallelogram we obtain $\angle CBP = \angle CP'P = \angle PDC$ as desired.

(b) (IMO 2006 shortlist) Since $\angle DCB + \angle CBA = 180^\circ$, the angles APB and CQD add up to 180°. Again, we want to reconstruct the cyclic quadrilateral. This time it is homothety that does the job.

Denote the intersection of AD and BC by E and consider homothety with center E which takes AB to DC. Then, for the image P' of P, we have $\angle DP'C = \angle APB$ and thus $DQCP'$ is cyclic, just as we intended.

Also, since the points K, L divide the segments AB, DC in the same ratio, line KL passes through E. Now we may conveniently erase K and L and leave a line through E only.

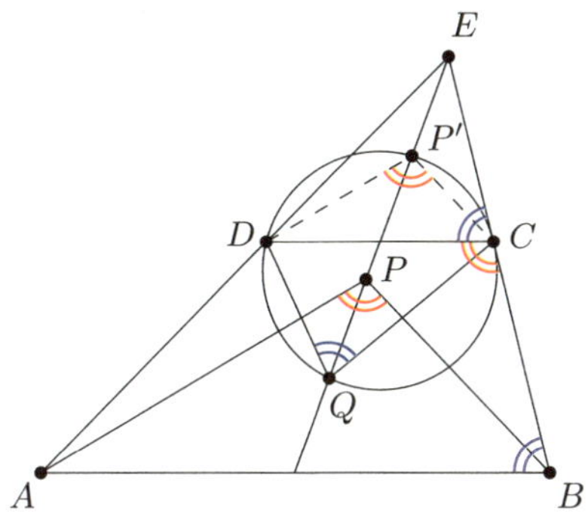

What remains is just some angle-chasing. From $\angle CQD = \angle CBA = \angle ECD$ we infer that BE is tangent to the circumcircle of $DQCP'$. Thus, the lines QC and $P'C$ are antiparallel in $\angle PEB$ and since by homothety $P'C \parallel PB$, we get that QC and PB are also antiparallel in $\angle PEB$ implying that P, Q, B, C are concyclic.

16. [Poland 2000] Let ABC be an isosceles triangle with base BC. Let P be a point inside the triangle ABC such that $\angle CBP = \angle ACP$. Denote by M the midpoint of the base BC. Show that $\angle BPM + \angle CPA = 180°$.

Proof. First, we address the constraint $\angle CBP = \angle ACP$. It implies that the circumcircle of triangle BCP, which we denote by ω, is tangent to the line AC. By symmetry in line AM, it is tangent to AB as well.

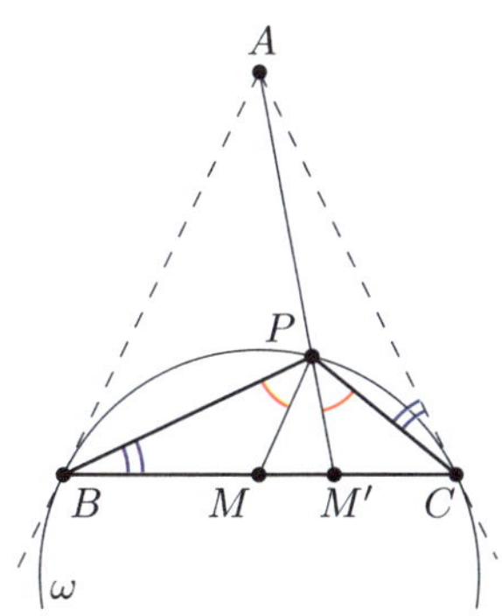

Now we focus on the triangle BCP. As A is the intersection of tangents to its circumcircle at the vertices B and C, line PA is its P-symmedian (see Introductory Problem 49). Thus, if we denote by M' the intersection of PA and BC, we obtain $\angle BPM = \angle M'PC$, which finishes the proof.

17. Let ABC be a non-right triangle with orthocenter H and circumcircle ω.

    (a) Let P be a point on ω. Prove that the reflections of P over the sides of the triangle ABC are collinear with H. Deduce that Simson line[3] of P with respect to triangle ABC bisects the segment PH.

    (b) Let ℓ be a line passing through H and denote by ℓ_a, ℓ_b, ℓ_c its reflections over the respective sides of the triangle ABC. Prove that ℓ_a, ℓ_b, ℓ_c pass through a common point on ω.

Proof.

(a) Denote the images of P in reflections over BC, CA, AB, by P_a, P_b, and P_c, respectively. We will prove only that P_b, P_c, and H are collinear, the rest will follow by analogous arguments.

The key idea is to introduce the images H_b and H_c of the orthocenter in reflections over AC and AB, respectively. Since both H_b and H_c lie on the circumcircle of triangle ABC (see Proposition 1.36), they are the natural link between the orthocenter and reflections in triangle sides.

Observe that triangles AHP_b and AH_bP are reflections of one another in line AC. In particular, they are congruent (but differently oriented). The same holds for triangles AHP_c and AH_cP. This should be enough to finish the problem by angle-chasing. Indeed, using oriented angles (to cover all possible positions of P) yields

$$\angle(AH, HP_b) = -\angle(AH_b, H_bP) = -\angle(AH_c, H_cP) = \angle(AH, HP_c),$$

where in the second equality we used that A, H_b, H_c, and P are concyclic. The conclusion follows (see Proposition 1.18).

As for the Simson line, consider homothety $\mathcal{H}(P, \frac{1}{2})$. It takes P_a, P_b, and P_c to the projections of P to BC, CA, AB, respectively, and hence it takes the line through P_a, P_b, and P_c to the Simson line of P with respect to triangle ABC. Since the line through P_a, P_b, and P_c passes through H, the Simson line of P with respect to triangle ABC passes through the midpoint of PH.

[3]For explanation see Proposition 1.44.

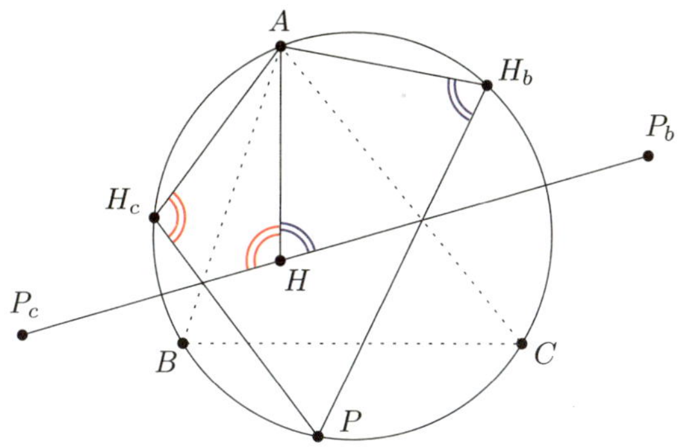

(b) (Anti-Steiner[4] point) Again, we only prove that the intersection X of non-parallel lines l_b and l_c lies on ω.

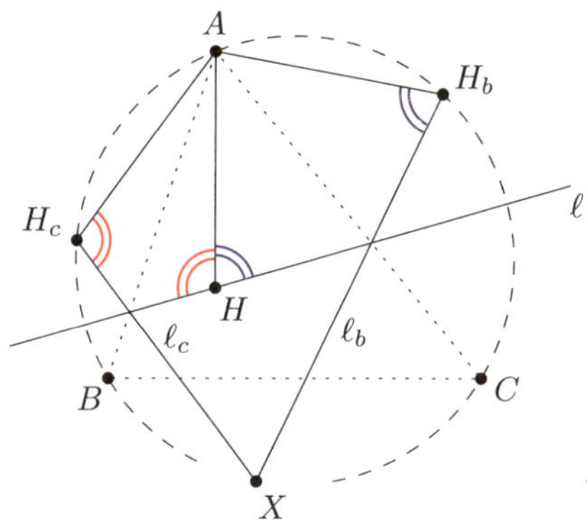

Note that l_b passes through H_b and l_c passes through H_c. Using the symmetries similarly as in (a), we again make use of directed angles:

$$\angle(XH_b, H_bA) = -\angle(\ell, HA) = \angle(XH_c, H_cA).$$

Thus points X, A, H_b, and H_c lie on one circle as desired.

18. Circles ω_a, ω_b are externally tangent at T and their common external tangent ℓ is tangent to them at A, B, respectively. Let ω be a circle inscribed in the curvilinear triangle ABT and denote by O its center and by r its radius. Prove that $OT \leq 3r$.

Proof. We invert about T with such radius, that ω is preserved (if in doubt, consult Introductory Problem 53) and superimpose the diagram with the original one.

[4]Jakob Steiner (1796—1863) was a Swiss mathematician who laid foundations of modern synthetic geometry.

In this inversion, circles ω_a, ω_b are mapped to parallel lines ω_a', ω_b' tangent to $\omega' = \omega$, and line ℓ is mapped to a circle ℓ' inscribed in the stripe formed by ω_a' and ω_b', tangent to ω and passing through T.

By now the result is apparent. Since both ℓ' and ω are inscribed in the same stripe, they are equal and thus denoting the point of contact of ω and ℓ' by X we have $OT \leq OX + XT \leq r + 2r = 3r$.

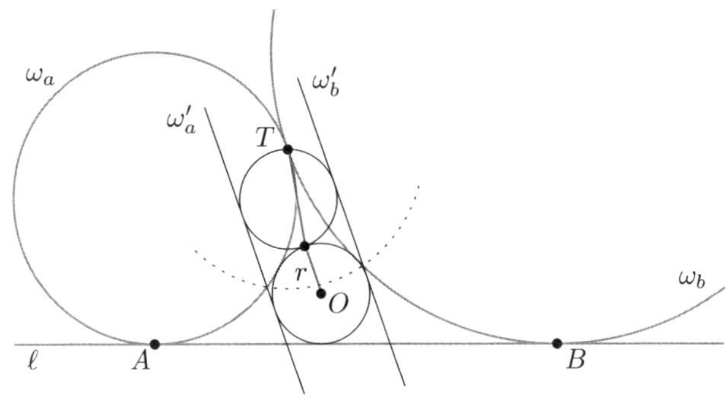

19. Let ABC be a triangle inscribed in circle ω and denote by R, r, r_a, r_b, r_c its circumradius, inradius, and the respective exradii.

(a) Denote by M the midpoint of the side BC and by N the midpoint of arc BC of ω containing vertex A. Prove that

$$MN = \frac{1}{2}(r_b + r_c).$$

(b) Prove that

$$r_a + r_b + r_c = 4 \cdot R + r.$$

(c) Let D, E, F be the midpoints of arcs BC, CA, AB of ω not containing vertices A, B, C, respectively. Prove that the perimeter of the hexagon $AFBDCE$ is at least $4(R + r)$.

Proof. Denote the incenter of triangle ABC by I and its respective excenters by I_a, I_b, I_c.

(a) Place BC horizontally. Since N is the midpoint of the segment I_bI_c (see the Big Picture – Proposition 1.42), the horizontal level of the point N is the average of the horizontal levels of the points I_b, I_c. But these are precisely the respective exradii which finishes the proof of the first part.

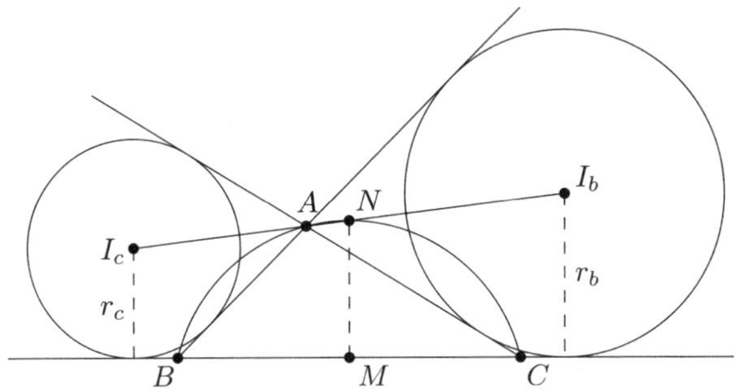

(b) Let D be the midpoint of arc BC of ω not containing vertex A. Then D is the midpoint of the segment II_a (again, recall the Big Picture) and as in the part (a) we conclude that $DM = \frac{1}{2}(r_a - r)$. As DN is the diameter of ω, summing this with the result of the first part we obtain the desired

$$r_a + r_b + r_c - r = 2 \cdot MN + 2 \cdot DM = 4 \cdot R.$$

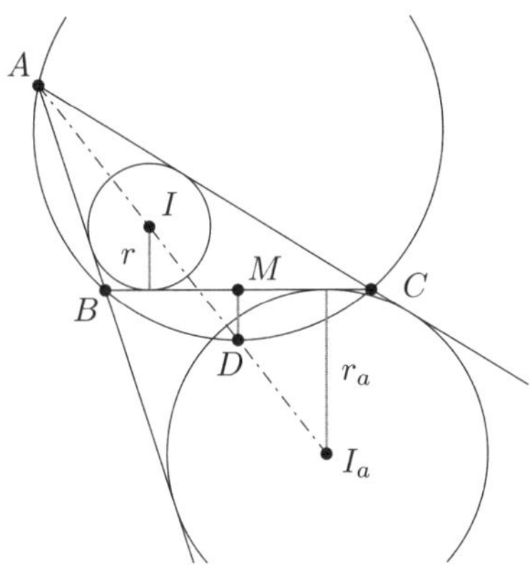

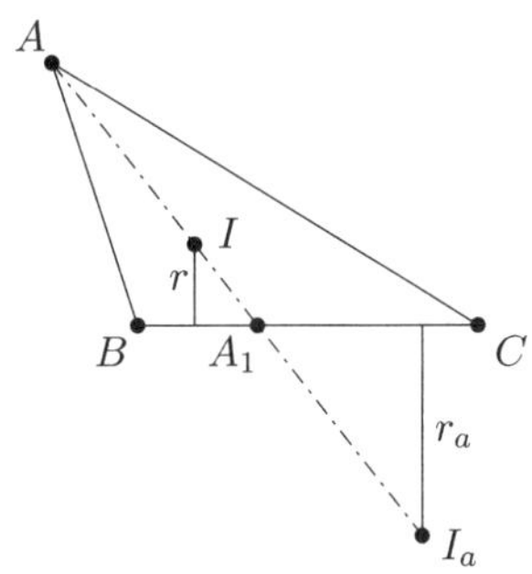

(c) (Mathematical Reflections, Michal Rolínek) Since $DB = DC = DI = DI_a$, the perimeter of $AFBDCE$ rewrites as

$$(BD + DC) + (CE + EA) + (AF + FB) = II_a + II_b + II_c.$$

By (b) we are to prove that this is at least $(4 \cdot R + r) + 3r = (r_a + r) + (r_b + r) + (r_c + r)$. A bit of wishful thinking now suggests we focus on much smaller diagram and try to prove $II_a \geq r_a + r$, since if we succeeded then the result would follow by adding symmetric inequalities. Fortunately, the mentioned inequality is not only true but also obvious.

Indeed, denoting by A_1 the foot of angle bisector by $\angle A$ we immediately have $IA_1 \geq r$ and $A_1 I_a \geq r_a$, and thus also $II_a \geq r_a + r$.

20. Circles ω_1, ω_2, and ω_3 are given in the plane, every one outside the others. Circle ω is tangent to them externally at A_1, A_2, A_3, respectively, and circle Ω is tangent to them internally at B_1, B_2, B_3, respectively. Prove that lines A_1B_1, A_2B_2, and A_3B_3 are concurrent.

First Proof. Proving concurrence of lines defined by tangency points of some circles should remind us of homothety.

Point A_1 is the center of negative homothety which maps ω to ω_1, and point B_1 is the center of positive homothety which maps ω_1 to Ω. Since performing the former homothety followed immediately by the latter one gives us the negative homothety which maps ω to Ω, line A_1B_1 passes through the center H^- of negative homothety between ω and Ω (see Lemma 1.31).

For exactly the same reason, lines A_2B_2 and A_3B_3 pass through H^- too. This finishes the proof.

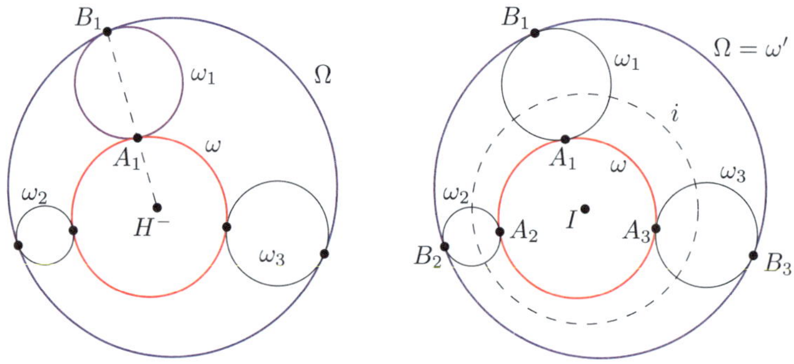

Second Proof. This time we handle the circles with the aid of inversion.

As in the Introductory Problem 53 we construct a circle i such that all three circles ω_1, ω_2, ω_3 are preserved in inversion about i. This inversion maps circle ω, which lies inside i, to a circle which lies outside i and is tangent to $\omega_1' = \omega_1$, $\omega_2' = \omega_2$, and $\omega_3' = \omega_3$. But there is only one such circle – namely Ω! Hence ω is mapped to Ω and in particular, points A_1, A_2, A_3 are mapped to B_1, B_2, B_3, respectively. Since any line through a point and its image in inversion passes through the center of that inversion, lines A_1B_1, A_2B_2, and A_3B_3 are concurrent at the center I of i.

21. [Kazakhstan 2012] Points K, L on the side BC of a triangle ABC satisfy $\angle BAK = \angle CAL < \frac{1}{2}\angle A$. Let ω_1 be any circle tangent to the

lines AB and AL, let ω_2 be any circle tangent to the lines AC and AK, and suppose that ω_1 and ω_2 intersect at P and Q. Prove that $\angle PAC = \angle QAB$.

Proof. Denote the intersections of ω_1 and ω_2 such that $AP < AQ$.

Points B, K, L, C are clearly mentioned in the problem for the notation purposes only. The diagram in fact consists of an angle (BAC), two isogonal lines in it (AK, AL), and two circles inscribed in the angles formed by these lines and the sides of the angle. In such setting, some sort of $\sqrt{bc}$-inversion is a must.

Denote the points of tangency of ω_1 with AB by T_1 and that of ω_2 with AC by T_2. Consider the transformation obtained by reflection about the bisector of angle BAC followed by inversion with center A and radius $\sqrt{AT_1 \cdot AT_2}$.

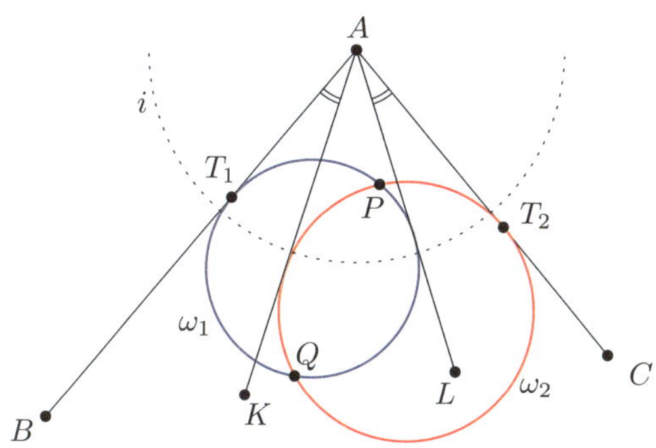

In such transformation, ω_1 is mapped to the circle inscribed in $\angle KAC$ tangent to line AC at the point with distance

$$\frac{r^2}{AT_1} = \frac{AT_1 \cdot AT_2}{AT_1} = AT_2$$

from A. Hence it is mapped to ω_2 and ω_2 is mapped to ω_1. Point P, being the intersection of ω_1 and ω_2 closer to A, is then mapped to the intersection of ω_2 and ω_1 further from A, i.e. to the point Q. Since point and its image in such transformation lie on isogonal lines, the result follows.

22. [All-Russian Olympiad 2011] An acute-angled triangle ABC is given. A circle passing through A and the triangle's circumcenter O intersects AB and AC at points P and Q, respectively. Prove that the orthocenter of the triangle POQ lies on the line BC.

First Proof. Denote the orthocenter by H.

If we manage to prove that quadrilaterals $BHOP$ and $CHOQ$ are cyclic, we will be instantly done as $\angle BHO + \angle OHC = \angle APO + \angle OQA = 180^\circ$. By symmetry, it suffices to prove the concyclicity of say $BHOP$ only.

The figure consists of the triangle ABC with its circumcenter O and the triangle POQ with its orthocenter H. These two parts are connected via cyclic quadrilateral $APOQ$. This guides us during the angle-chasing

$$\angle OHP = 90^\circ - \angle HPQ = \angle PQO = \angle PAO = \angle OBA$$

which shows that $BHOP$ is cyclic and we may end the proof here.

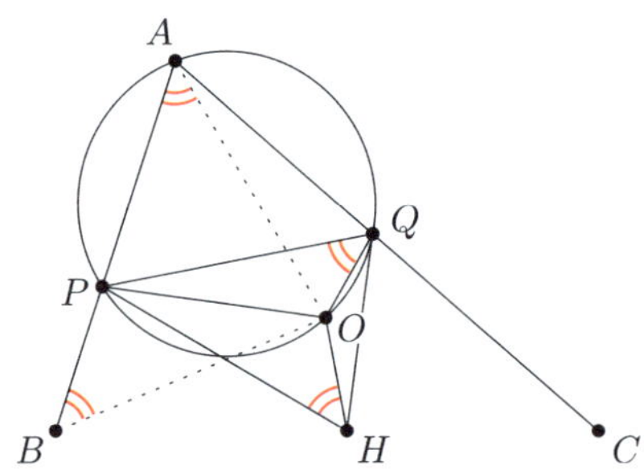

Second Proof. Denote by K, L, M the midpoints of the sides BC, CA, AB. First, we will check the statement for $P = M$ and $Q = L$ (note that $AMOL$ is cyclic) and for the general case we will use dynamic argument.

We have already seen that O is the orthocenter in triangle KLM (see Introductory Problem 23(b)) which means K is the orthocenter of triangle OLM (see Lemma 1.34). Since $K \in BC$, the case $P = M$ and $Q = L$ is done.

Now consider points $P \neq M$, $Q \neq L$ on the sides AB, AC such that $APOQ$ is cyclic.

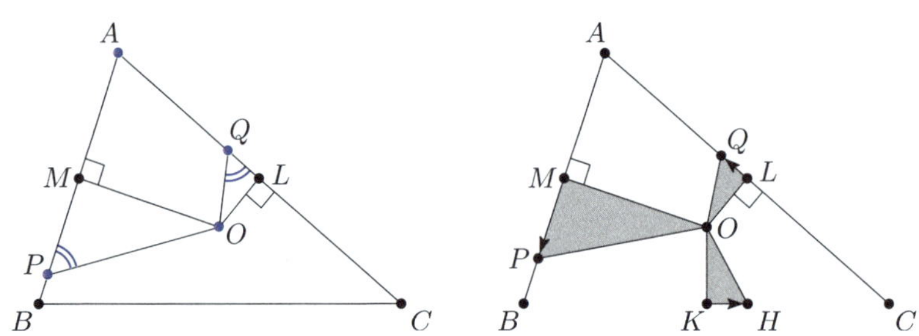

Since $\angle OQL = 180^\circ - \angle AQO = \angle OPM$, right triangles OLQ and OMP are similar (AA). We consider the spiral similarity $\mathcal{S}$ centered at O which takes L to Q and M to P. Note that denoting the orthocenter of triangle OPQ by H, all we need is $\angle HKO = 90^\circ$.

Since $\mathcal{S}$ takes triangle OLM to triangle OQP it takes the orthocenter of triangle OLM (i.e. K) to the orthocenter of triangle OPQ (i.e. H). Thus, $\triangle OKH \sim \triangle OMP \sim \triangle OLQ$ and indeed $\angle OKH = \angle OMP = 90^\circ$.

23. [All-Russian Olympiad 2002] Let O be the circumcenter of a triangle ABC. Points M and N are chosen on the sides AB and AC, respectively, so that $\angle NOM = \angle A$. Prove that the perimeter of triangle MAN is not less than the length of the side BC.

Proof. This is going to be tricky! Our strategy will be to rearrange the sides of triangle AMN so that they form a broken line. Then it should be easier to compare its total length with BC. The presence of the circumcenter (a point equidistant from A, B, and C) suggests using rotation.

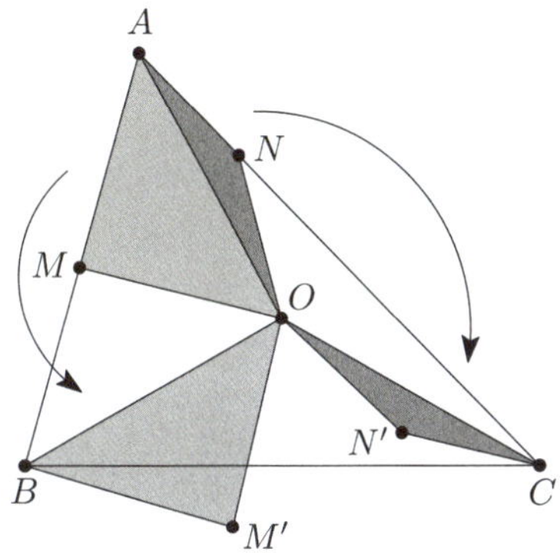

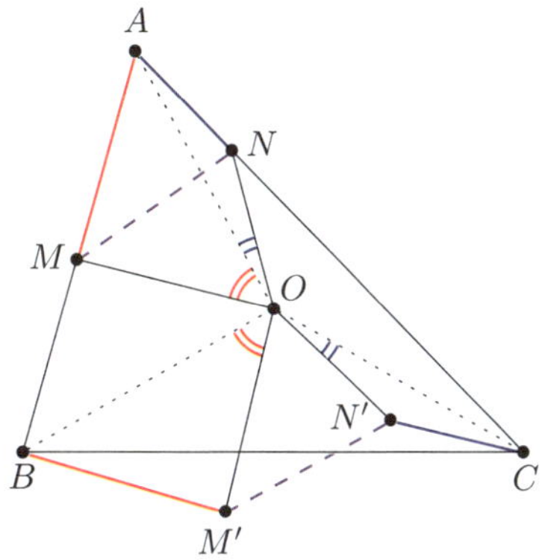

First, we consider rotation with center O which takes A to B and apply this rotation to triangle AOM. The image of M will be denoted as M'. Similarly, we consider rotation with center O which takes A to C and apply it to triangle AON to obtain a new point N'. Since rotation preserves distances, we have $BM' = AM$ and $CN' = AN$. Now, we wish to prove $M'N' = MN$, since then the conclusion would follow (a straight line is the shortest distance from B to C). As we have $OM = OM'$ and $ON = ON'$ we only need to prove $\angle M'ON' = \angle NOM$ to ensure the SAS congruence of triangles MON and $M'ON'$. But this is easy, because $\angle BOC = 2\angle A$ (central angle) and

$$\angle M'ON' = \angle BOC - (\angle BOM' + \angle N'OC) = 2\angle A - \angle NOM = \angle A,$$

and we may conclude.

24. [Sharygin Geometry Olympiad 2005] Let ABC be a scalene triangle with orthocenter H and incenter I. Line ℓ_a is perpendicular to the bisector of $\angle A$ and passes through the midpoint of BC. Lines ℓ_b and ℓ_c are defined analogously. Show that the circumcenter O_1 of triangle formed by these lines lies on the line IH.

Proof. We aim to relate point O_1 to some triangle centers of triangle ABC. First, we get rid of the midpoints. Denote by G, the centroid of triangle ABC and recall that homothety $\mathcal{H}_1(G, -2)$ takes the midpoint of BC to A and thus line ℓ_a goes to a line ℓ'_a through A perpendicular to the internal angle bisector. In other words, ℓ'_a is the external angle bisector. Since the same holds for ℓ_b and ℓ_c, the triangle formed by the images has the excenters I_a, I_b, and I_c of triangle ABC as vertices. Also O_1 goes to O_2, the circumcenter of triangle $I_aI_bI_c$.

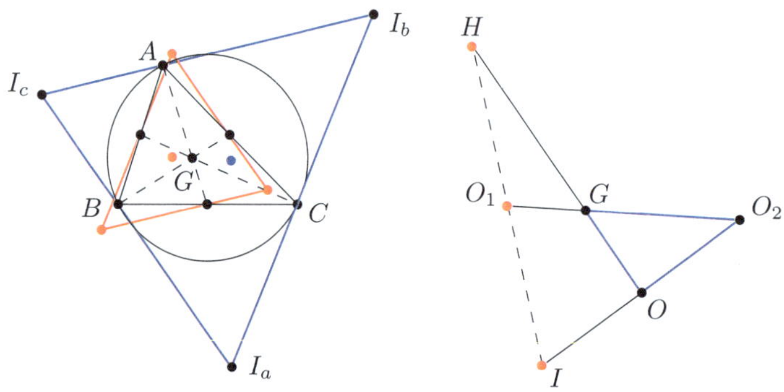

In order to connect O_2 with triangle ABC we use the Big Picture (see Proposition 1.42). Recall that the circumcircle of triangle ABC centered at point O is the nine-point circle of triangle $I_aI_bI_c$ and that I is the orthocenter in triangle $I_aI_bI_c$. Hence as in the proof of the nine-point circle (see Theorem 1.37) homothety $\mathcal{H}_2(I, \frac{1}{2})$ takes O_2 to O.

Finally, we found a construction of O_1 from the triangle centers of triangle ABC and we can draw a diagram depicting it. Since H, G, and O lie on the Euler Line (see Example 1.3) in a known ratio, we have enough information to conclude. Either we recognize a familiar diagram with triangle HIO_2 and its centroid G or we can mindlessly verify collinearity of O_1, I, and H using Menelaus Theorem in triangle GOO_2. Indeed, as we have

$$\frac{IO}{IO_2} \cdot \frac{O_1O_2}{O_1G} \cdot \frac{HG}{HO} = \frac{1}{2} \cdot \frac{3}{1} \cdot \frac{2}{3} = 1,$$

the proof is over.

25. Let ω_a, ω_b be two circles that are externally tangent at T and internally tangent to circle ω at A, B, respectively. Let S be one of the intersections of the common tangent of ω_a, ω_b at T with ω. Line AS intersects ω_a again at C and BS intersects ω_b again at D. Line AB intersects ω_a again at E and ω_b again at F. Prove that lines ST, CE, DF are concurrent.

Proof. Since ST is the radical axis of ω_a, ω_b, by the Radical Lemma it suffices to prove that the points C, D, E, F lie on a single circle (see Proposition 1.23).

By Introductory Problem 45, line CD is the common external tangent of ω_a, ω_b.

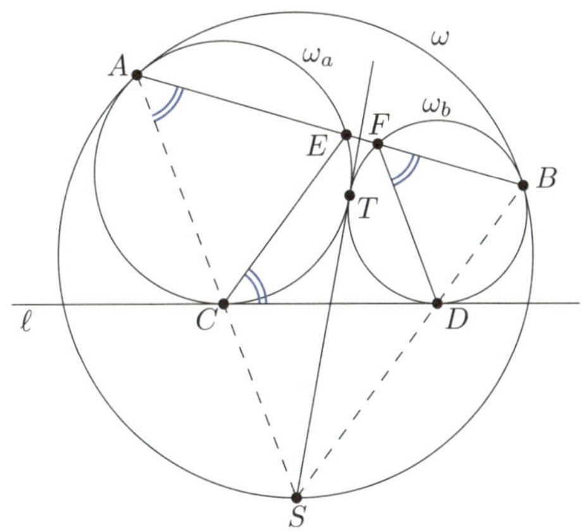

Hence $\angle DCE = \angle CAE$. But since the homothety centered at B which takes ω to ω_b maps AS to FD, we have $AS \parallel FD$ and $\angle CAE = \angle SAB = \angle DFB$ which ensures that $CDFE$ is cyclic as desired.

26. Shortest paths.

(a) Let ℓ be a line and A, B two points on the same side of it. For what point $L \in \ell$ is $AL + LB$ minimal?

(b) Let ABC be an acute-angled triangle. Among all the triangles DEF with vertices D, E, F on the sides BC, CA, AB, respectively, one has minimal perimeter. Find which one.

Solution of (a). In order to estimate the length of the broken line we aim to straighten it.

Let B' be the reflection of B about ℓ. Then for any point X on the line ℓ we have $AX + XB = AX + XB' \geq AB'$ and the equality occurs if

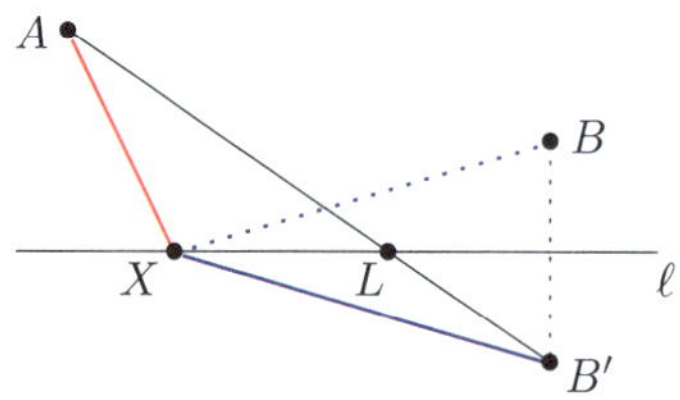

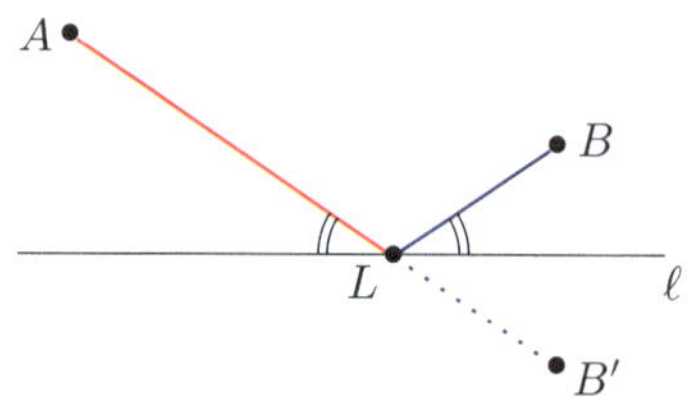

$X \in AB'$. Hence the point L we are looking for is the intersection of ℓ with AB'.

First Solution of (b). (Fagnano[5]'s problem) If D, E, F are the points on the respective sides of triangle ABC such that the perimeter of triangle DEF is the minimal possible then by (a) the segments DE, DF form the same angle with BC and likewise for the other sides. In other words, lines BC, CA, AB are the respective external angle bisector in triangle DEF implying that A, B, C are its respective excenters.

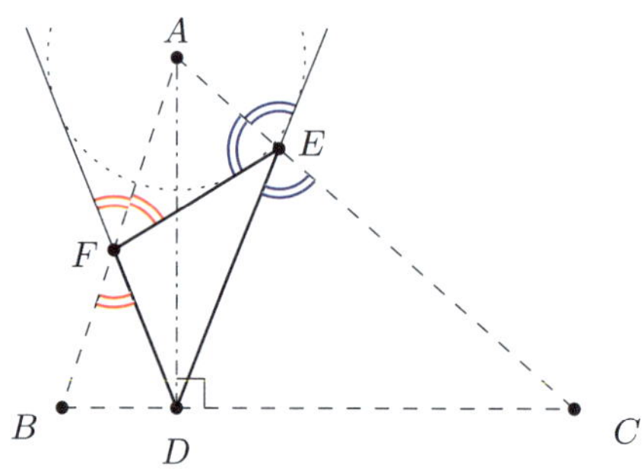

Being the D-excenter of triangle DEF, point A lies on the bisector of angle FDE. Since the internal and external angle bisectors of $\angle FDE$ are perpendicular, point D is the foot of A-altitude in triangle ABC. Likewise we learn that E and F are the feet of the other altitudes.

The triangle with minimal perimeter is the one formed by the feet of the altitudes.

Second Solution of (b). Guided by the first part, fix D on the side BC and let D', D'' be the reflections of D about the sides AB, AC, respectively. Then

$$DF + FE + ED = D'F + FE + ED'' \geq D'D''$$

and hence we aim to find such point D on BC that the distance $D'D''$ is minimal.

[5]Giovanni Francesco Fagnano dei Toschi (1715–1797) was an Italian archpriest with extensive interest in mathematics.

As D varies along BC, its reflections about the sides AB, AC vary along segments BC', $B'C$, where triangles ABC', ACB' are the reflections of the original triangle ABC about its sides AB, AC. Moreover, $BD' = BD = B'D''$, so we may temporarily simplify the diagram again – now it consists of two congruent triangles $AC'B$, ACB' with D', D'' being corresponding points on their sides $C'B$, CB'.

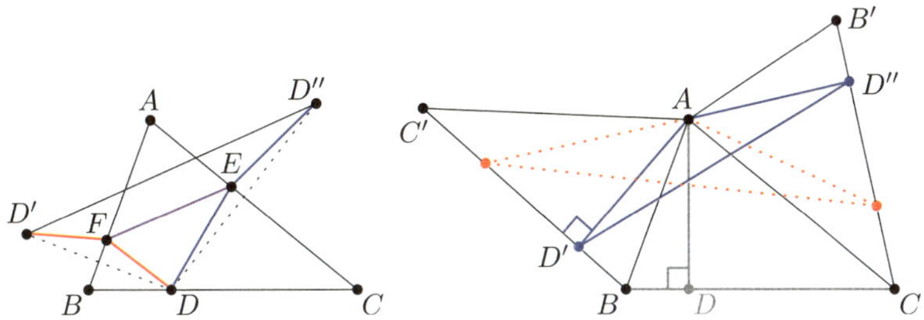

The spiral similarity centered at A which maps triangle $AC'B$ to triangle ACB' (in fact it is rotation) maps D' to D''. Hence all the triangles $AD'D''$ have the same shape and in order to minimize $D'D''$ we may minimize AD' instead. The point on $C'B$ closest to A is the projection of A onto $C'B$ which (back in triangle ABC) corresponds to D being the foot of altitude from A.

By the same reasoning we conclude that E, F are the feet of altitudes too.

Remark. Note that the second solution of (b) does not require the hypothesis that a triangle with minimal perimeter actually exists. If we wanted to remove this hypothesis also from the first solution, we would need to verify that there is no sequence of triangles with decreasing perimeters that tends to a degenerate case with one of D, E and F at a vertex.

27. [Based on IMO 1992 shortlist] Circles ω_1, ω_2 inscribed in a given circular sector with endpoints A, B are externally tangent at T. Denote by ℓ their common internal tangent.

    (a) Prove that ℓ passes through a fixed point independent of the position of ω_1, ω_2.

    (b) Let C be the intersection of ℓ with arc AB. Prove that T is the incenter of triangle ABC.

First Proof. Without loss of generality assume AB is horizontal and the circular sector is "above" it. The given arc AB determines a circle. Denote it by ω.

(a) We will prove that the fixed point is the midpoint M of arc AB of ω lying "below" AB.

Recall that common internal tangent is the radical axis of ω_1 and ω_2. Thus it suffices to prove $p(M, \omega_1) = p(M, \omega_2)$.

Denote by T_1, T_2, K, L the points of tangency of ω_1 and ω, ω_2 and ω, ω_1 and AB, and ω_2 and AB.

As K is the "bottom" point of the circle ω_1, a homothety centered at T_1 that sends ω_1 to ω maps K to M. Hence the points T_1, K, M are collinear and similarly, T_2, L, M are collinear.

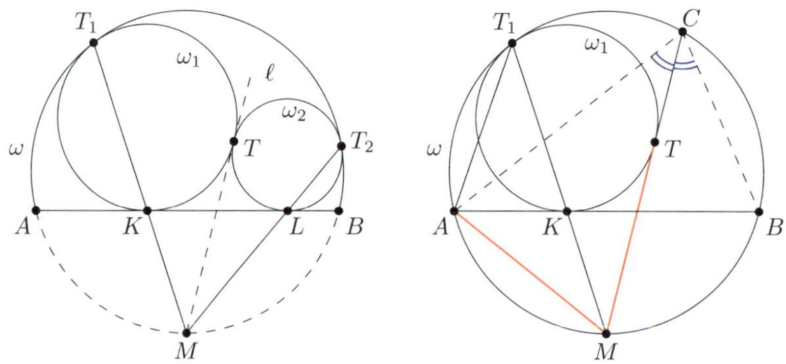

Now we are left to prove $MK \cdot MT_1 = ML \cdot MT_2$ which is true by the Shooting Lemma since both sides are equal to MA^2 (see Proposition 1.40(a)).

(b) As CT passes through M, it is the bisector of angle ACB. By the alternative definition of the incenter, it suffices to prove $MT = MA$ (see Proposition 1.39), which is straightforward, since

$$MT^2 = p(M, \omega_1) = MK \cdot MT_1 = MA^2.$$

Second Proof. Let ω be the circle containing arc AB and let ℓ meet ω at C and M with C on arc AB. Draw ℓ vertically and perform an inversion with respect to T. Denote images under this inversion with primes.

The circles ω_1 and ω_2 and the line ℓ become three vertical lines ω_1', ω_2' and ℓ' with ℓ' between the other two. The line AB becomes a circle Γ meeting ℓ' at T and tangent to ω_1' and ω_2'. The circle ω becomes a circle ω' tangent to ω_1' and ω_2' with T in its interior. The intersections of ω'

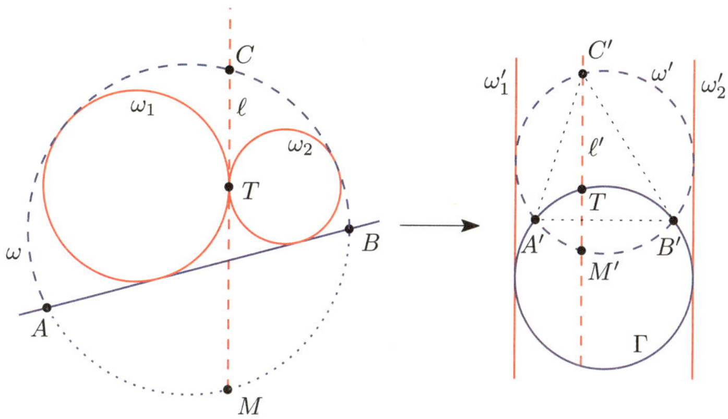

with Γ are A' and B'. The intersections of ℓ' with ω' are C' and M' with M' inside Γ.

Now by symmetry $A'B'$ is horizontal and M' is the reflection of T across $A'B'$. Hence T is the orthocenter of triangle $A'B'C'$ (see Proposition 1.36). Further this triangle is acute since its orthocenter is in its interior. Using the result of Introductory Problem 44 (and the fact that a second inversion about T will reverse the first), we see that T is the incenter of triangle ABC. This solves (b). From this (a) follows immediately since the fact that CT bisects $\angle ACB$ implies that M is the midpoint of the arc AB of ω not in the given circular segment, independent of the positions of ω_1 and ω_2.

28. [IMO 2005] Let $ABCD$ be a fixed convex quadrilateral with $BC = DA$ and BC not parallel to DA. Let two variable points E and F lie on the sides BC and DA, respectively, and satisfy $BE = DF$. The lines AC and BD meet at P, the lines BD and EF meet at Q, the lines EF and AC meet at R. Prove that the circumcircles of the triangles PQR, as E and F vary, have a common point other than P.

Proof. The most natural way to employ $BC = DA$ and $BE = DF$ is to consider rotation $\mathcal{R}$ which sends B to D and C to A (and hence also E to F). Denote the center of such rotation by S.

Since rotation is a special case of spiral similarity, its center S is the second intersection of the circumcircles of the triangles BCP and DAP (see Proposition 1.47). But in our case, $\mathcal{R}$ also sends BE to DF and EC to FA so it also lies on the circumcircles of the triangles BEQ, DFQ, ECR and FAR!

With so many properties it is not hard to guess and prove that S is the point we are looking for. For instance, if we make use of cyclic quadrilaterals $BCPS$ and $ECRS$ we conclude by

$$\angle(SR, RQ) \equiv \angle(SR, RE) = \angle(SC, CE) \equiv \angle(SC, CB) = \angle(SP.PB) \equiv$$
$$\equiv \angle(SP, PQ),$$

where we used directed angles in order to cover all possible cases.

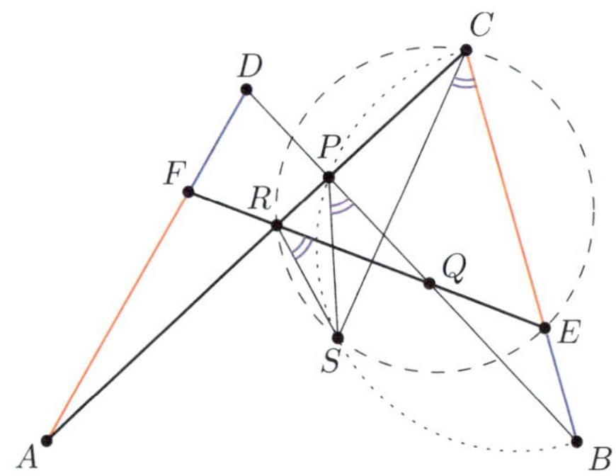

29. [All-Russian Olympiad 1995] Let $ABCD$ be a quadrilateral inscribed in a semicircle ω with diameter AB and center O. Lines CD and AB intersect at M. Let K be the second point of intersection of the circumcircles of triangles AOD and BOC. Prove that $\angle MKO = 90°$.

Proof. Consider inversion about ω and denote by M', K' the images of M and K. It suffices to prove $\angle OM'K' = 90°$ (see Proposition 1.51).

Clearly, points A, B, C, D are preserved in such inversion. The circumcircles of triangles AOD and BOC are mapped to the lines AD and BC so $K' = AD \cap BC$.

Line CD is mapped to the circumcircle of triangle COD (denote it by Γ) and line AB is mapped to itself so M' is the second intersection of Γ and AB.

Let us focus on triangle ABK' with altitudes AC and BD. As O is the midpoint of AB, Γ is the nine-point circle of this triangle (see Theorem 1.37) and hence M' is the foot of altitude from K' to AB. We may conclude.

Remark. The assertion remains valid even if AB is an arbitrary chord of circle ω. The interested reader is encouraged to prove this claim. (It can be done with angle-chasing only, if one guesses and exploits certain cyclic quadrilateral.)

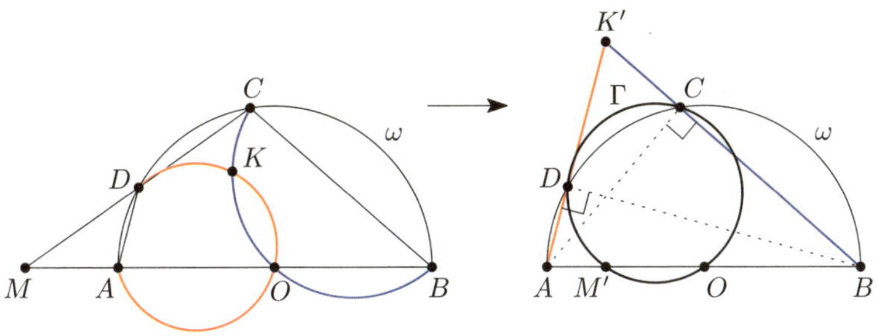

30. [Poland 2006] Let AB be a segment and C its midpoint. Circle ω_1 which passes through A and C intersects circle ω_2 which passes through B and C at two different points C and D. Point P is the midpoint of arc AD of circle ω_1 which does not contain C. Similarly, point Q is the midpoint of arc BD of circle ω_2 which does not contain C. Prove that $PQ \perp CD$.

Proof. We will show that CP and CQ have equal projections onto CD, which ensures $PQ \perp CD$.

Focus on the left half of the diagram only and note that since CP is the angle bisector of $\angle ACD$ (see Proposition 1.38(b)), we are dealing with a very standard configuration.

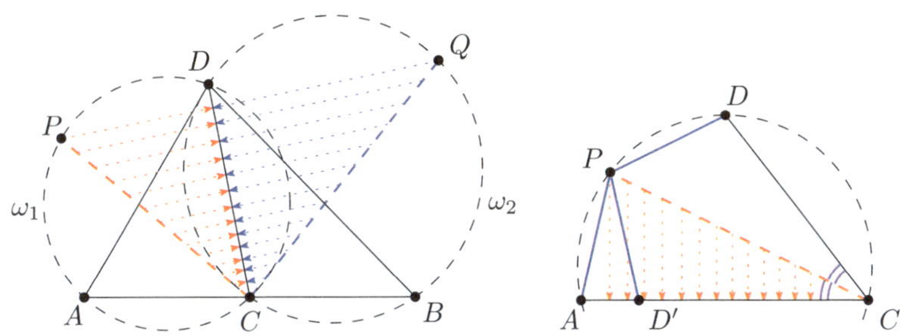

Among many possible ways to proceed we choose a fast (but a little tricky) one. Denote by D' the reflection of D in the angle bisector CP. Of course, $D' \in AB$ and $PD' = PD = PA$ (P is the midpoint of arc AD). Placing AC horizontally helps us realize that P then lies "above" the midpoint of AD', which implies that the projection of CP onto CA equals $\frac{1}{2}(CA + CD') = \frac{1}{2}(CA + CD)$. As CP is the angle bisector, the projection onto CD is the same. Should D' coincide with A, $ACDP$ is a cyclic kite with diameter CP and we get the same conclusion.

Likewise we find that the projection of CQ onto CD or CB equals $\frac{1}{2}(CB + CD)$ and we may conclude.

31. [Mathematical Reflections, Michal Rolínek] Let BC be a fixed chord of the circle ω with radius R and let A vary on the major arc BC of ω forming an acute triangle ABC with $\angle A \neq 60^\circ$ and orthocenter H.

    (a) Show that the mirror images H' of H over the A-angle bisector run along a circle.

    (b) Show that the projections X of H on the A-angle bisector also run along a circle.

Proof of (a). Observe that AH' is isogonal to AH in $\angle BAC$, therefore (see Proposition 1.17), A, H', and O are collinear, where O is the circumcenter of triangle ABC. Moreover, $AH' = AH = 2R|\cos\angle A|$ (see Proposition 1.35(f)), which is fixed. Hence $OH' = |AO - AH'| = R|1 - 2\cos\angle A|$ is also fixed, implying that H' moves along a circle with center O and (nonzero) radius $R|1 - 2\cos\angle A|$.

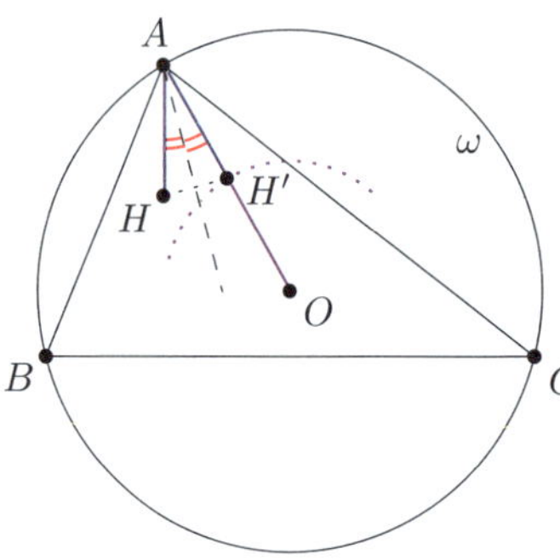

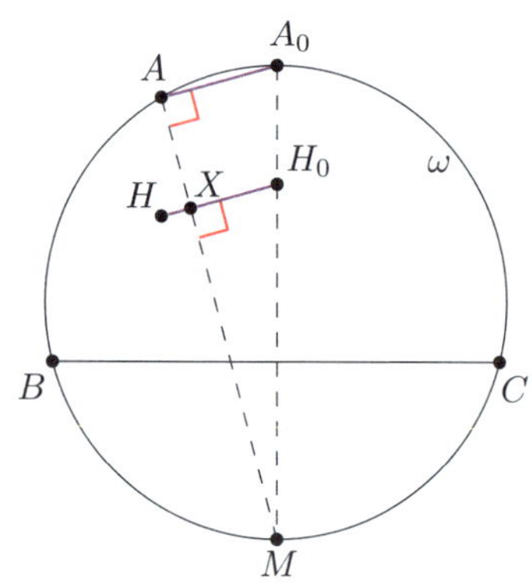

First Proof of (b). Denote by A_0 and M the midpoints of the major and minor arcs BC of ω, respectively. Also, let H_0 be the orthocenter of A_0BC and observe that for $A = A_0$, point X coincides with H_0. We will prove that X lies on a circle with diameter MH_0.

First, observe that AX and A_0H_0 both meet ω at M. Further, in Introductory Problem 16, we have proved that $HH_0 \parallel AA_0$ and since MA_0 is a diameter of ω, we have $AA_0 \perp AM$ and so $HH_0 \perp AM$, yielding $X \in HH_0$ and also $\angle MXH_0 = 90^\circ$. This proves our assertion.

Second Proof of (b). Recall that H also describes a circle, in fact the reflection of ω in BC (see Proposition 1.36). Since both H' (from part (a)) and H trace a circle with the same relative speed (namely the speed of A along ω), the Averaging Principle immediately yields that so do their midpoints X.

32. [Sharygin Geometry Olympiad 2012] In acute triangle ABC inscribed in circle ω, let A' be the projection of A onto BC and B', C' the projections of A' onto AC, AB, respectively. Line $B'C'$ intersects ω at X and Y and line AA' intersects ω for the second time at D. Prove that A' is the incenter of triangle XYD.

Proof. First we prove that DA bisects $\angle XDY$. Denote the circumcenter of triangle ABC by O and recall that AO and AA' are isogonal in $\angle BAC$ (see Proposition 1.17).

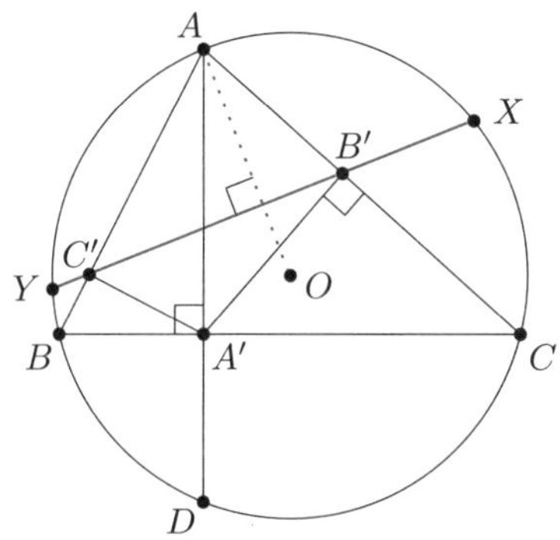

As $\angle AB'A' = 90° = \angle AC'A'$, line AA' passes through the circumcenter of triangle $AB'C'$, and hence AO (being isogonal to it also in $\angle B'AC'$) is perpendicular to $B'C'$. A line perpendicular to a chord of a circle through the center of that circle is its perpendicular bisector so A is the midpoint of arc XY of ω. As a consequence, DA bisects $\angle XDY$ (if in doubt, see Proposition 1.38(b)).

Now it suffices to prove $AA' = AX$ (see the alternative definitions of the incenter – Proposition 1.39(b)). This might seem a bit hopeless at first, but as $AX^2 = AB' \cdot AC$ (see Shooting Lemma 1.40(a)), we quickly get rid of X and are left to prove $AB' \cdot AC = AA'^2$ in right triangle $AA'C$.

If the last equality is not obvious yet, consult Introductory Problem 2.

33. [China TST 2006] Given a triangle ABC, let B_1, B_2, and C_1, C_2 be points on the sides AB and AC, respectively, such that $BB_1/BB_2 = CC_1/CC_2$. Prove that the orthocenters of triangles ABC, AB_1C_1, and AB_2C_2 are collinear.

Proof. We choose to define the orthocenters as intersections of B and C altitudes and look at the problem dynamically.

Imagine a pair of lines ℓ_b and ℓ_c such that $\ell_b \perp AC$ and $\ell_c \perp AB$, which start their motion with $B \in \ell_b$ and $C \in \ell_c$ and move uniformly until $B_2 \in \ell_b$ and $C_2 \in \ell_c$, one of the positions then being $B_1 \in \ell_b$ and $C_1 \in \ell_c$ (since points B_1 and C_1 divide BB_2 and CC_2 in the same ratio). It suffices to prove that the intersections of ℓ_b and ℓ_c move along a line, which sounds more than reasonable.

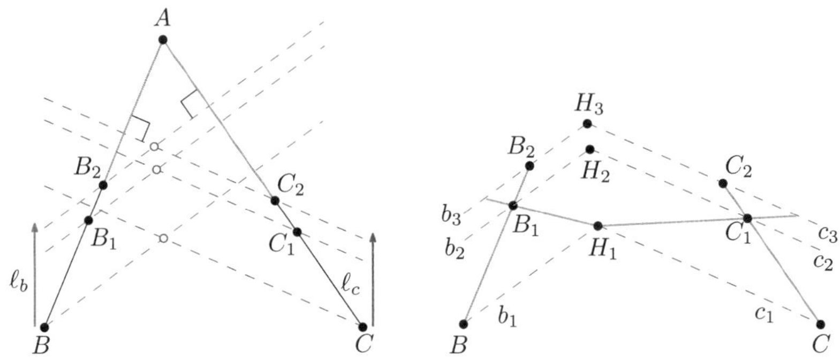

Label the three positions of ℓ_b and ℓ_c as b_1, b_2, b_3, and c_1, c_2, c_3 and their three intersections as H_1, H_2, and H_3. Now just observe that homothety centered at H_1 which sends b_2 to b_3 has factor $BB_2/BB_1 = CC_2/CC_1$ and thus sends c_2 to c_3. Hence it maps H_2 to H_3, which proves the desired collinearity.

34. [All-Russian Olympiad 2009] Let ABC be a scalene triangle. The angle bisector of $\angle A$ intersects the side BC at D and the circumcircle Ω of triangle ABC at A and E. Circle ω with diameter DE cuts Ω again at F. Prove that AF is the symmedian[6] of triangle ABC.

First Proof. First observe that the midpoint M of BC lies on ω as $\angle DME = 90°$. Now consider $\sqrt{bc}$-inversion. Since the endpoints D, E of diameter of ω are interchanged, the circle itself remains intact. But since $\sqrt{bc}$-inversion swaps BC and Ω, point M clearly goes to F, implying that lines AF and AM are isogonal in $\angle A$. We may conclude.

Second Proof. As in the first proof, M is the midpoint of BC and lies on ω. Also, let N be antipodal to E on Ω (hence E, M, and N are collinear). Since $\angle EFD = 90°$, the ray FD intersects Ω again at N. Finally, as $\angle DMN = 90°$ and $\angle EAN = 90°$ (EN is diameter of Ω), the quadrilateral $DMNA$ is cyclic. Now we are ready to show the isogonality of AF and AM by angle-chasing:

$$\angle FAE = \angle FNE \equiv \angle DNM = \angle DAM.$$

[6]For explanation see Introductory Problem 49.

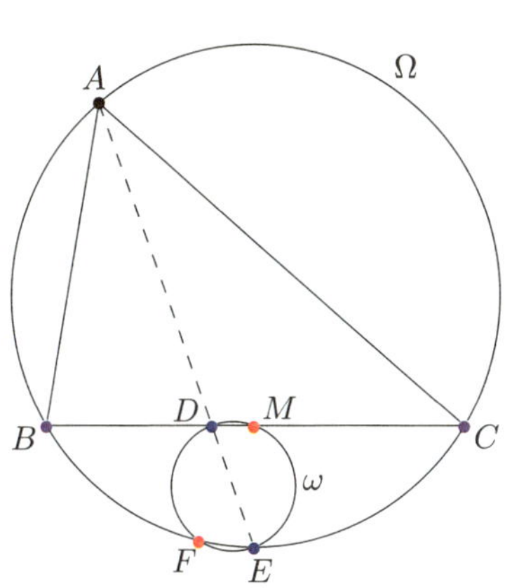

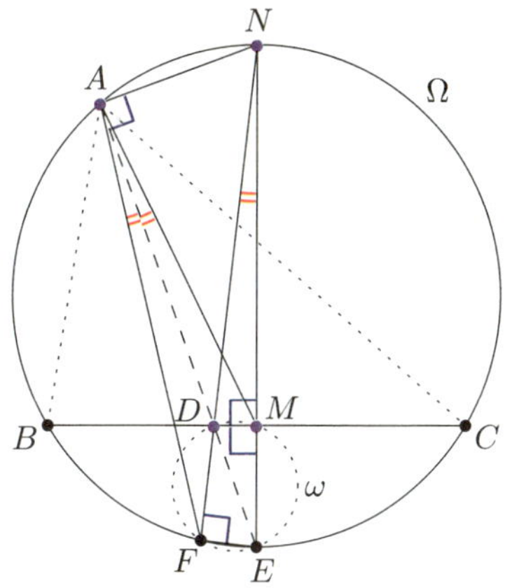

35. [Baltic Way 2006] Let ABC be a triangle, let K be the midpoint of the side AB and L the midpoint of the side AC. Let P be the second intersection of the circumcircles of triangles ABL and AKC. Let Q be the second intersection of AP and the circumcircle of triangle AKL. Prove that $2AP = 3AQ$.

Proof. Seeing the busy point A, we decide to straighten things up a bit following the idea of $\sqrt{bc}$-inversion. We slightly adjust this technique by changing the radius of inversion to $\sqrt{\frac{1}{2}bc}$, since then $K' = C$ and $L' = B$.

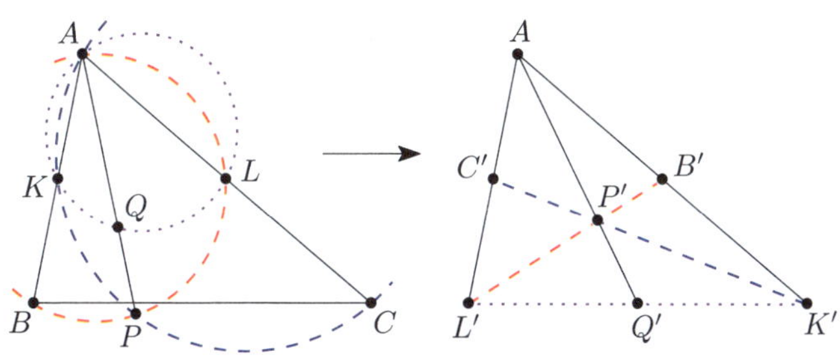

Then P' is the intersection of the medians $B'L'$ and $C'K'$ in triangle $AL'K'$ i.e. the centroid. Also Q' is the intersection of AP' with $K'L'$ which is the midpoint of $K'L'$. Since medians divide each other in the ratio $2:1$, we have $3AP' = 2AQ'$. In the original picture this rewrites as $3/AP = 2/AQ$ and we may conclude.

Remark. Here combining the inversion with reflection is not really necessary. On the other hand, it lends extra perspective by showing that AP is a symmedian (for explanation see Introductory Problem 49) in triangle ABC.

36. An angle of fixed magnitude φ revolves about its fixed vertex A and meets a fixed line ℓ at points B and C. Prove that the circumcircles of triangles ABC are all tangent to a fixed circle.

Proof. We invert about A. Now ℓ transforms to a circle ℓ' with $A \in \ell'$, the angle still revolves about A and $B', C' \in \ell'$. We aim to prove that lines $B'C'$ are tangent to a fixed circle.

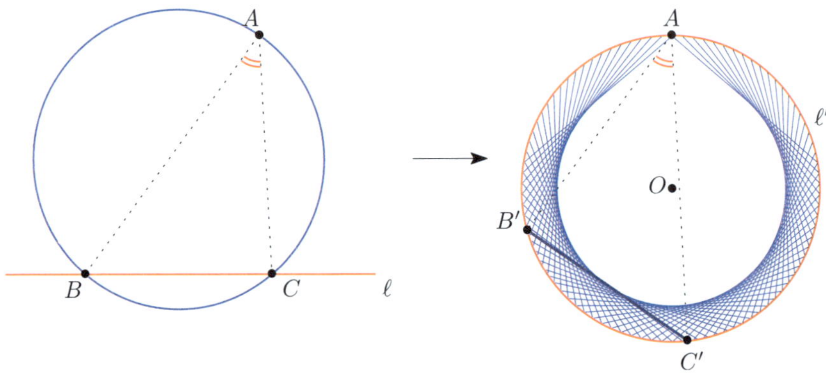

But all the possible segments $B'C'$ are chords of ℓ' with the same corresponding inscribed angle φ. Therefore, the segments $B'C'$ are all equal and thus they keep fixed distance d from the center O of ℓ'. In other words, the lines $B'C'$ are all tangent to the circle with center O and radius d.

37. [Iran 2011] Let ABC be a triangle and denote its circumcircle centered at O by ω. Points M and N lie on the sides AB and AC, respectively. The circumcircle of triangle AMN intersects ω for the second time at Q. Let P be the intersection point of MN and BC. Prove that PQ is tangent to ω if and only if $OM = ON$.

Proof. Without loss of generality assume that Q lies on the arc AB of ω not containing C and observe that it is the Miquel point of the quadrilateral $BCNM$ (see Theorem 1.49). Hence the quadrilaterals $PBMQ$ and $PCNQ$ are cyclic too.

First we assume that PQ is tangent to ω. We angle-chase.

Since $PBMQ$ is cyclic, $\angle PQB = \angle PMB = \angle NMA$ holds. On the other hand, as PQ is tangent to ω we may write $\angle PQB = \angle QAB$. Thus, $QA \parallel MN$.

As $QMNA$ is a cyclic trapezoid, it is isosceles and the perpendicular bisectors of QA and MN coincide. However, O lies on the bisector of QA so it also lies on the bisector of MN and $OM = ON$ as required.

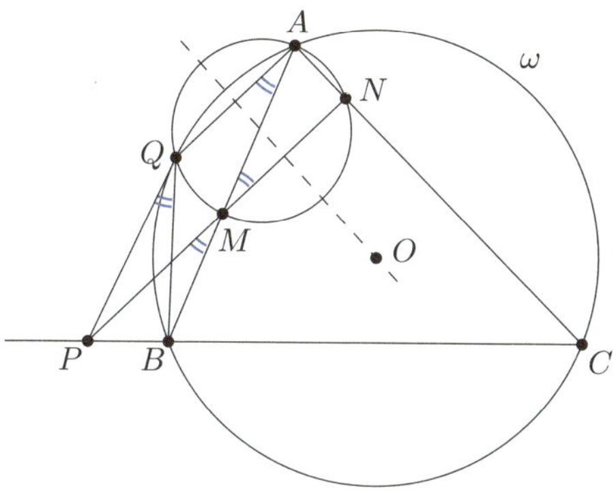

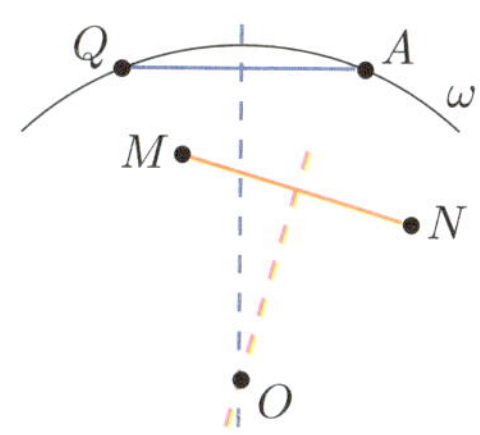

Now we prove the "if" part. Assume $OM = ON$.

Observe that both the perpendicular bisectors of QA and MN pass through O. If they did not coincide, O would have to be the circumcenter of (cyclic) $QMNA$. But that is impossible, since the points M, N lie inside ω and $OM < OA$. Hence the segments QA and MN share the perpendicular bisectors which implies that $QMNA$ is an isosceles trapezoid.

Finally, similarly as in the first part, we angle-chase $\angle PQB = \angle PMB \equiv \angle NMA = \angle QAB$ and conclude that PQ is tangent to ω.

38. [USA TST 2000] Let $ABCD$ be a cyclic quadrilateral. The projections of the intersection of its diagonals P to the sides AB and CD are E, F, respectively. Show that the line EF is perpendicular to the line through the midpoints K and L of the sides of BC and DA, respectively.

Proof. Here we present a spectacular application of spiral similarity. We would like to use the Averaging Principle on the two similar triangles ABP and DCP ($ABCD$ is cyclic!) but we can't since the similarity is indirect. We fix this by reflecting P over CD to get P' and directly similar triangles ABP and DCP'. Then their average which is triangle LKF has also their shape.

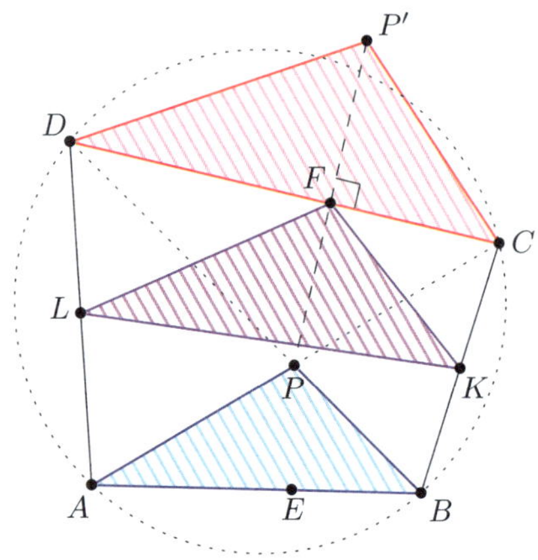

Likewise we show that triangle LEK has this very shape too. The quadrilateral $LEKF$ is then formed by two congruent triangles glued together along KL, therefore it is a kite and we may conclude.

39. [IMO 2010] Given a triangle ABC with incenter I and circumcircle Γ, let AI intersect Γ again at D. Let E be a point on the arc BDC, and F a point on the segment BC, such that $\angle BAF = \angle EAC < \frac{1}{2}\angle BAC$. If G is the midpoint of IF, prove that lines EI and DG intersect on Γ.

First Proof. Points E, F lie on isogonal lines with respect to $\angle BAC$, one of them on the circumcircle of triangle ABC, the other one on the side BC. What does it mean? Yes, they are images of one another in $\sqrt{bc}$-inversion!

Since the incenter I is present in the diagram we recall that its image in $\sqrt{bc}$-inversion is the A-excenter I_a (see Introductory Problem 33) and draw it too.

Now

$$AI \cdot AI_a = bc = AE \cdot AF \quad \text{and} \quad \angle IAE = \angle FAI_a,$$

hence $\triangle IAE \sim \triangle FAI_a$ (SAS) and in particular $\angle AEI = \angle AI_aF$. Furthermore, as we know from the Big Picture (see Proposition 1.42), point D is the midpoint of II_a and thus DG is the midline in triangle FII_a and $\angle AI_aF = \angle ADG$.

Equal angles AEI and ADG are both inscribed in Γ, hence they intercept the same arc implying that EI and DG intersect at Γ.

Second Proof. Let EI intersect Γ for the second time at X. Equivalently, we may prove that DX bisects FI. Let G' be the intersection.

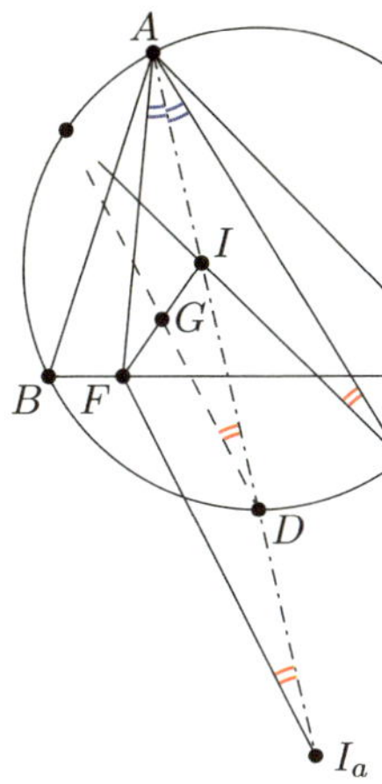

Since $\angle DXE = \angle DAE = \angle FAD$, if we denote the intersection of AF and XD by H then $HIAX$ is cyclic. In other words, line HI is antiparallel to XA with respect to the angle $\angle XDA$.

But since D is the midpoint of arc BC, line BC is also antiparallel to XA in angle $\angle ADX$ (see Proposition 1.40(c)). Hence $HI \parallel BC$.

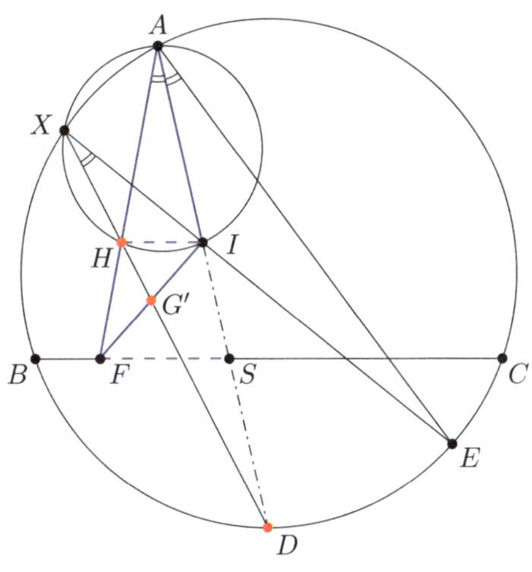

Now we will prove that $FG'/G'I = 1$. Let us focus on triangle AFI and collinear points H, G', D on its sides. Menelaus' Theorem implies

$$\frac{AH}{HF} \cdot \frac{FG'}{G'I} \cdot \frac{ID}{DA} = 1 \quad \text{and hence} \quad \frac{FG'}{G'I} = \frac{HF}{AH} \cdot \frac{AD}{DI}.$$

Since $HI \parallel BC$, the first fraction rewrites as $HF/AH = SI/IA$ ($\triangle AHI \sim \triangle AFS$), where S denotes the intersection of BC and AD. Thus the whole problem is reduced to the metric identity concerning the points on the angle bisector only, namely

$$SI \cdot AD = DI \cdot IA,$$

which is proved in Introductory Problem 33(c).

40. [Czech-Polish-Slovak Match 2008] Let $ABCDE$ be a regular pentagon. Find the minimum possible value of

$$\frac{PA + PB}{PC + PD + PE}$$

where P is any point in the plane.

Solution. We may assume $AB = 1$ and let d denote the length of the diagonal in $ABCDE$. We shall use the Ptolemy's Inequality (see Theorem 1.46) multiple times. Indeed, if we apply it for (possibly degenerate or self-intersecting) quadrilaterals $APBC$, $APBD$, $APBE$ (with vertices in this order!), we obtain

$$\begin{aligned} PA \cdot 1 + PB \cdot d &\geq 1 \cdot PC, \\ PA \cdot d + PB \cdot d &\geq 1 \cdot PD, \\ PA \cdot d + PB \cdot 1 &\geq 1 \cdot PE. \end{aligned}$$

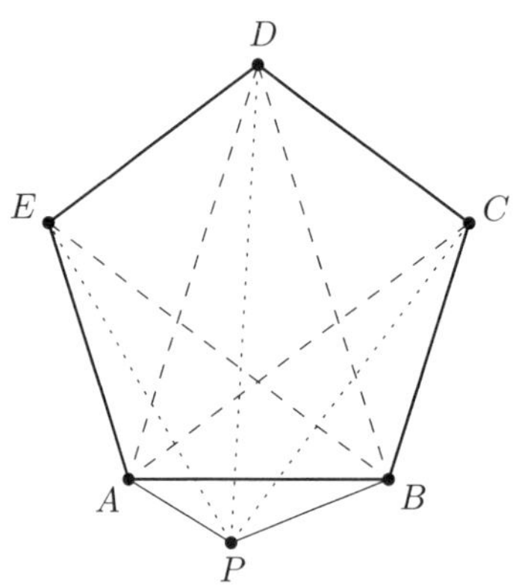

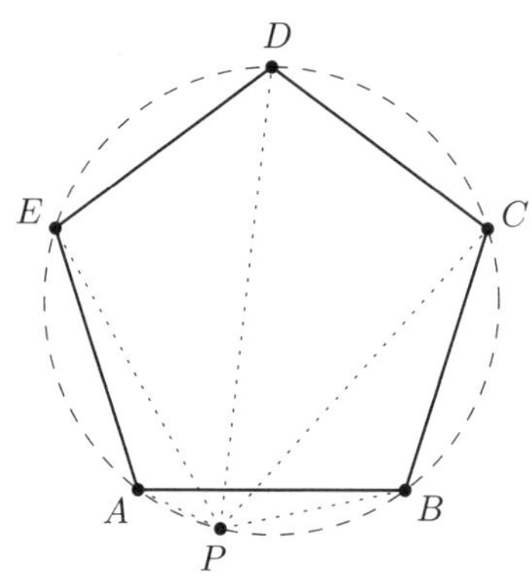

Addition yields

$$(PA + PB)(1 + 2d) \geq PC + PD + PE,$$

hence

$$\frac{PA + PB}{PC + PD + PE} \geq \frac{1}{1 + 2d}.$$

Since this value is attained if P lies on the minor arc AB of the circumcircle of $ABCDE$, it is the sought-after minimum.

It remains to calculate d, which should not be too difficult. For example, we may again use Ptolemy's Inequality (in equality case) for $ABCD$ to see that $1+d=d^2$. Hence

$$d=\frac{1+\sqrt{5}}{2}, \quad \text{and} \quad \frac{1}{1+2d}=\sqrt{5}-2,$$

which is our final answer.

41. [Poland 2012] Let ABC be an A-isosceles triangle inscribed in circle Ω. Arbitrary circles ω_b, ω_c inscribed in the minor circular segments AC, AB of Ω are tangent to Ω at B', C', respectively. One of the common external tangents of ω_b and ω_c intersects the sides AC, AB at P, Q, respectively. Prove that lines $B'P$ and $C'Q$ intersect on the angle bisector of $\angle BAC$.

Proof. The key here is to figure out a way to deal with line $B'P$ (and similarly $C'Q$). Since B' is the center of positive homothety which maps Ω to ω_b, we aim to interpret P as a center of another homothety hoping to exploit Lemma 1.31.

Let ω be the incircle of triangle APQ. As P is the center of negative homothety which maps ω_b to ω, the mentioned lemma ensures that line $B'P$ passes through the center H^- of negative homothety between Ω and ω.

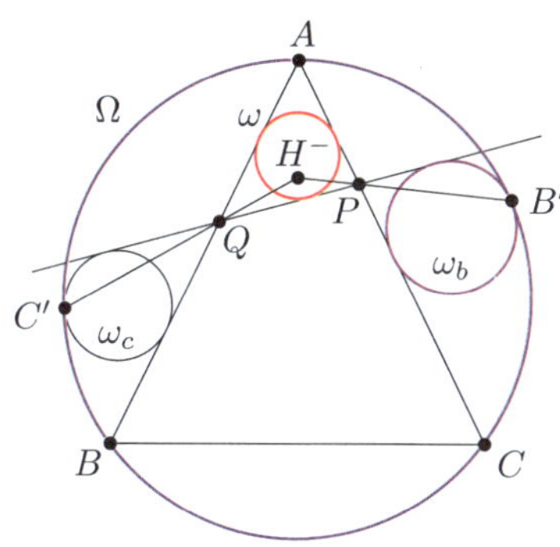

Similarly, we argue that $C'P$ also passes through H^-. Hence the intersection of $B'P$ and $C'Q$ is H^-. The conclusion now follows since the angle bisector of BAC is the common line of symmetry of both ω and Ω (recall that $AB=AC$).

42. [USAMO 2001] Let ABC be a triangle and let ω be its incircle. Denote by D_1 and E_1 the points where ω is tangent to the sides BC and AC, respectively. Denote by D_2 and E_2 the points on sides BC and AC, respectively, such that $CD_2 = BD_1$ and $CE_2 = AE_1$, and denote by P the point of intersection of segments AD_2 and BE_2. Circle ω intersects segment AD_2 at two points, the closer of which to the vertex A is denoted by Q. Prove that $AQ = D_2P$.

Proof. Using the standard notation, $CE_2 = AE_1 = x$ and $CD_1 = BD_2 = z$. We will show that

$$\frac{D_2P}{PA} = \frac{AQ}{PA}.$$

The first ratio is readily found from Menelaus' Theorem applied for triangle ACD_2 and line BP. We have

$$\frac{D_2P}{PA} \cdot \frac{AE_2}{E_2C} \cdot \frac{CB}{BD_2} = 1, \quad \text{hence} \quad \frac{D_2P}{PA} = \frac{x}{z} \cdot \frac{z}{a} = \frac{x}{a}.$$

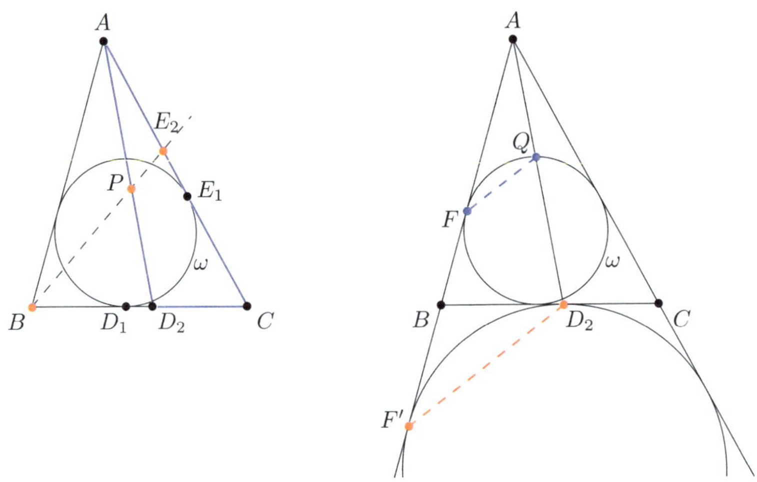

On the other hand, D_2 is the point of tangency of the A-excircle (call it ω_a) with BC (recall Proposition 1.7(c)). Now we denote by F, F' the points of tangency of line AB with ω and ω_a, respectively. Then the homothety centered at A which takes ω to ω_a, also takes F to F' and Q to D_2. Thus $\triangle AFQ \sim \triangle AF'D_2$ and the ratios yield

$$\frac{AQ}{QD_2} = \frac{AF}{FF'} = \frac{AF}{AF' - AF} = \frac{x}{s - x} = \frac{x}{a},$$

where the penultimate equality follows from Proposition 1.7(b). We may conclude.

43. [USAMO 2008] Let ABC be an acute, scalene triangle, and let M, N, and P be the midpoints of BC, CA, and AB, respectively. Let the perpendicular bisectors of AB and AC intersect ray AM in points D and E, respectively, and let lines BD and CE intersect in point F, inside triangle ABC. Prove that points A, N, F, and P all lie on one circle.

First Proof. First, we note that triangles BDA and CEA are isosceles. Let $\angle BAM = \delta$ and $\angle MAC = \varphi$. Summing angles in quadrilateral $BFCA$ gives $\angle BFC = 2\delta + 2\varphi = 2\angle A$, which means that F lies on the circumcircle of triangle BCO, where O is the circumcenter of triangle ABC. Once O is in the diagram, we realize it suffices to prove $\angle OFA = 90^\circ$, since the circumcircle of ANP has diameter AO. Now we can erase points N and P.

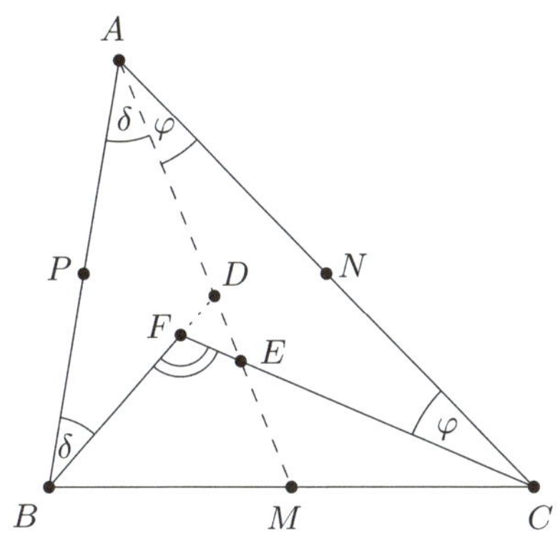

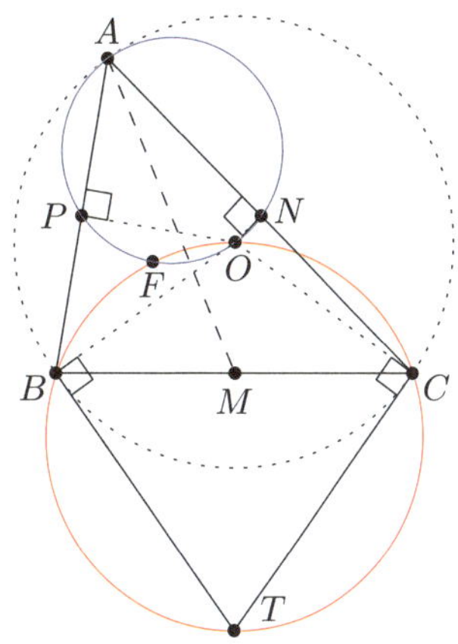

Looking at circle BOC, we may as well decide to prove that A and F are collinear with the point T which is diametrically opposite to O. The vital step is to observe that T is the intersection of tangents to the circumcircle of triangle ABC at B and C. Proving collinearity of A, F, and T is thus equivalent to proving that AF is a symmedian in triangle ABC (see Introductory Problem 49).

We will compare angles CTF and CTA. Since AT is a symmedian, we have

$$\angle CTA = (180^\circ - \angle ACT) - \angle TAC = \angle B - \delta$$

and the cyclic quadrilateral $TBFC$ gives

$$\angle CTF = \angle CBF = \angle B - \delta.$$

Then points A, F, and T are indeed collinear and we may conclude.

Second Proof. As in the first proof we start by observing that triangles BDA and CEA are isosceles and that $\angle BFC = 2\angle A$. Then the trick

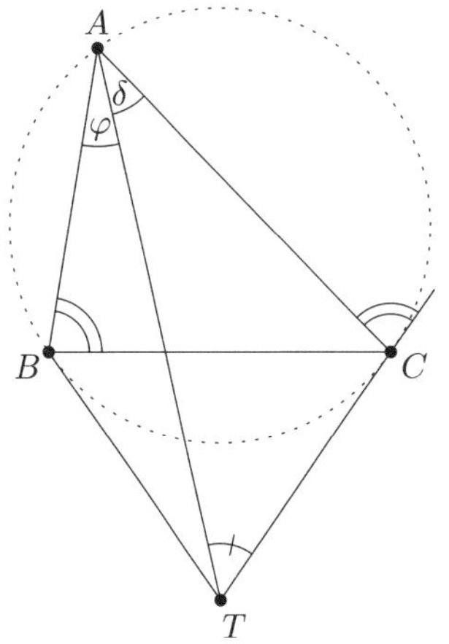

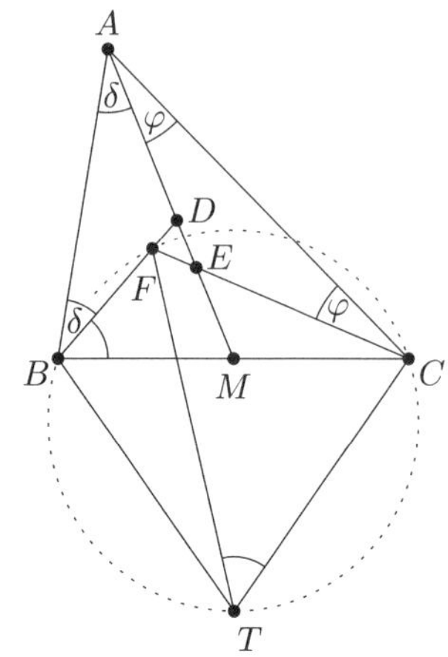

is to use the Law of Sines and show that $\angle AFB = \angle CFA$. Since B, F, and C are not collinear it suffices to show the angles have the same sine.

From triangles BFA and CFA we learn (keeping the notation from the first proof)

$$\sin\angle AFB = \sin\delta \cdot \frac{AB}{AF}, \qquad \sin\angle CFA = \sin\varphi \cdot \frac{AC}{AF}$$

so to prove the angles are equal, we only need $AB \cdot \sin\delta = AC \cdot \sin\varphi$. But this follows from the Law of Sines applied in triangles ABM and BCM:

$$AB \cdot \sin\delta = MB \cdot \sin\angle AMB = MC \cdot \sin\angle CMA = AC \cdot \sin\varphi.$$

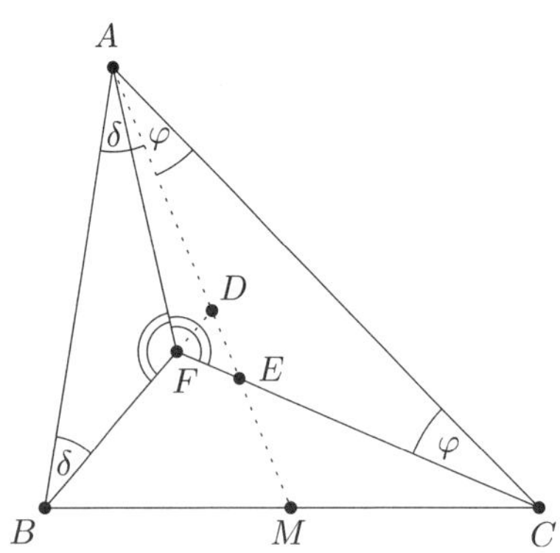

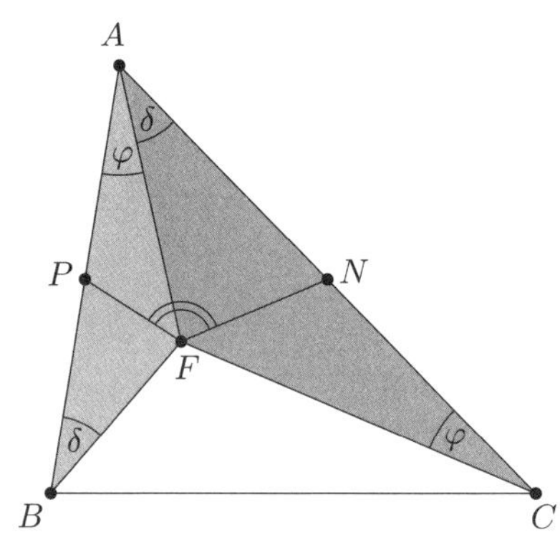

Since $\angle AFB + \angle CFA = 360^\circ - 2\angle A$, we know that $\angle AFB = \angle CFA = 180^\circ - \angle A$. From here we also deduce $\angle BAF = \varphi$ and $\angle FAC = \delta$.

Then $\triangle AFC \sim \triangle BFA$ and moreover, spiral similarity $\mathcal{S}(F, k, 180^\circ - \angle A)$ takes triangle AFC to triangle BFA for a suitable choice of k.

Thus, it also takes N to P, implying that $\angle NFP = 180^\circ - \angle A$, which means that $ANFP$ is indeed cyclic.

Third Proof. In this solution, we will also work with the circumcenter O (and assume $AB < AC$). Since both N and P lie on the circle with diameter AO it suffices to show $\angle AFO = 90^\circ$.

Now note that the perpendicular bisectors (in triangle ABC) PD and NE are external angle bisectors in triangle DFE. Therefore O is the F-excenter of this triangle and in particular, it lies on the internal bisector of $\angle EFD$. This is good news as it allows yet another reformulation of the problem. The line AF ought to be perpendicular to the internal angle bisector FO, in other words it ought to be the bisector of the external angle EFD. We will prove just that.

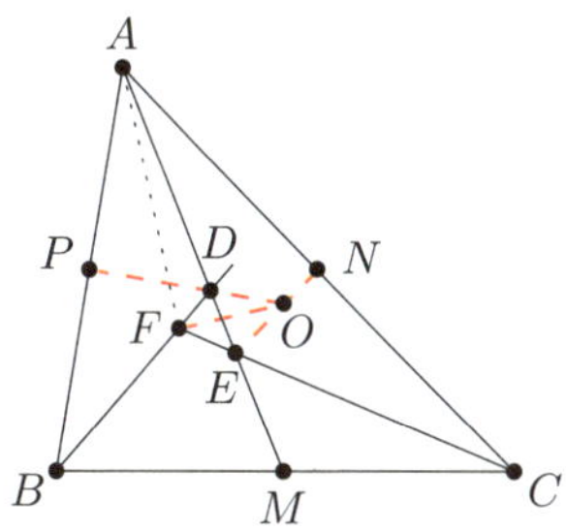

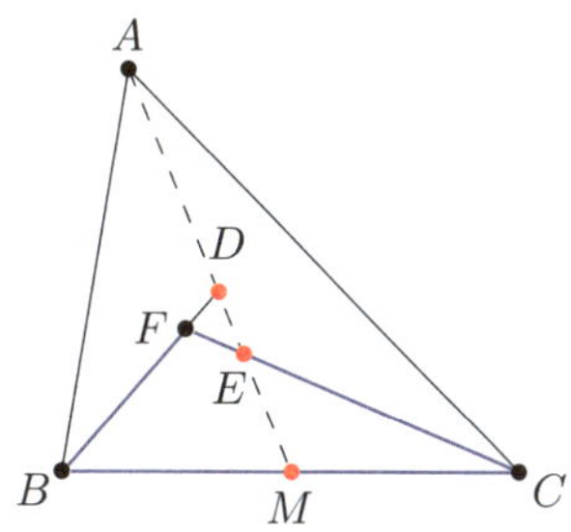

In Introductory Problem 27 we have seen that the foot of the external bisector is equivalently described in terms of ratios. The sought equality is

$$\frac{AD}{AE} = \frac{FD}{FE}$$

and due to isosceles triangles ADB and AEC, it can be further rewritten as

$$\frac{BD}{CE} = \frac{FD}{FE}.$$

The final blow is delivered by Menelaus' Theorem for triangle BCF and collinear points E, D, M which gives

$$\frac{BM}{MC}\frac{CE}{EF}\frac{FD}{DB} = 1.$$

As $BM = MC$, we obtain the desired equality and can conclude the solution.

44. [Balkan MO 2009] Let MN be a line parallel to the side BC of a triangle ABC, with M on the side AB and N on the side AC. The lines BN and

CM meet at point P. The circumcircles of triangles BMP and CNP meet at two distinct points P and Q. Prove that $\angle BAQ = \angle CAP$.

First Proof. First we recognize a familiar part of the diagram. Since Q is the second intersection of the circumcircles of triangles BMP and CNP, it is the Miquel point (see Theorem 1.49) of the quadrilateral $AMPN$ and hence it also lies on the circumcircles of the triangles ABN and ACM.

This suggests inverting about A (by far the most "busy" point around). But with what radius? As $MN \parallel BC$, we have

$$\frac{AM}{AB} = \frac{AN}{AC} \quad \text{or} \quad AM \cdot AC = AN \cdot AB.$$

Guided by the properties of $\sqrt{bc}$-inversion we invert about A with radius $\sqrt{AM \cdot AC} = \sqrt{AN \cdot AB}$ and reflect the result about the angle bisector of angle BAC.

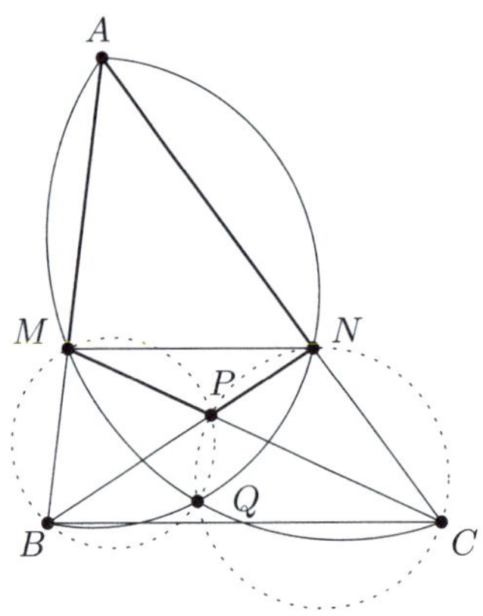

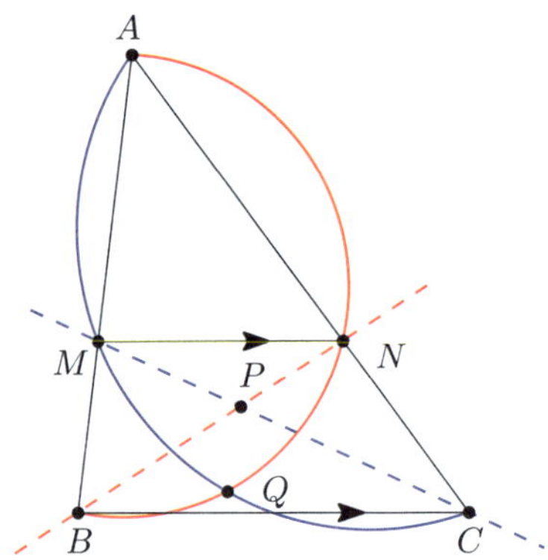

In such transformation, points M and C are interchanged and so are the points N and B. Hence the circumcircle of triangle AMC is mapped to the line MC and the circumcircle of triangle ANB is mapped to the line NB. As a result, point Q is mapped to P and thus $\angle BAQ = \angle CAP$.

Second Proof. First, we recall that a simple homothety argument (see Example 1.2) shows that P lies on the A-median of triangle ABC. The problem then asks to prove that Q lies on the corresponding symmedian (for explanation see Introductory Problem 49).

We will prove it using a characterization of symmedian points from Introductory Problem 49(c). We are left to prove the following: If Q_b and Q_c are the projections of Q on the sides AC and AB, then

$$\frac{QQ_b}{QQ_c} = \frac{AC}{AB}.$$

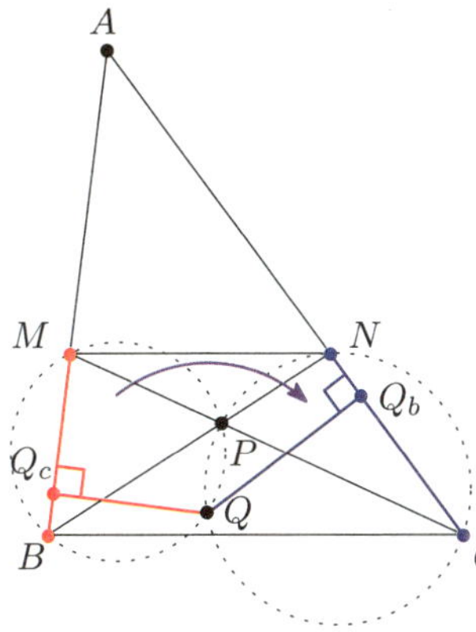

By construction, Q is the center of spiral similarity $\mathcal{S}(S, k, \varphi)$ taking BM to NC. Naturally, $\mathcal{S}$ also takes Q_c to Q_b. All in all, we have

$$\frac{QQ_b}{QQ_c} = k = \frac{NC}{BM}.$$

And that's all, since straightforward application of similar triangles AMN and ABC gives the remaining

$$\frac{NC}{BM} = \frac{AC}{AB}.$$

Not bad right?

45. [IMO 1998 shortlist] Let $ABCDEF$ be a convex hexagon such that $\angle B + \angle D + \angle F = 360°$ and

$$\frac{AB}{BC} \cdot \frac{CD}{DE} \cdot \frac{EF}{FA} = 1.$$

Prove that

$$\frac{BC}{CA} \cdot \frac{AE}{EF} \cdot \frac{FD}{DB} = 1.$$

Proof. We have to find a way to employ both the conditions simultaneously. The first one suggests bringing the angles by B, D and F together. In fact, we will glue together three triangles similar to triangles CDE, EFA, and ABC, respectively.

Looking at the desired condition, we see that among B, D, and F, it is point D that has a special role (it appears as endpoint of two diagonals). That's why we choose D to have a special role in our construction.

Let X be the point such that $\triangle EDX \sim \triangle EFA$ (directly). Then

$$\angle CDX = 360^\circ - \angle D - \angle F = \angle B \quad \text{and} \quad DX = FA \cdot \frac{ED}{EF} = BA \cdot \frac{CD}{CB}.$$

Thus, the triangles CDX and CBA are also similar (SAS).

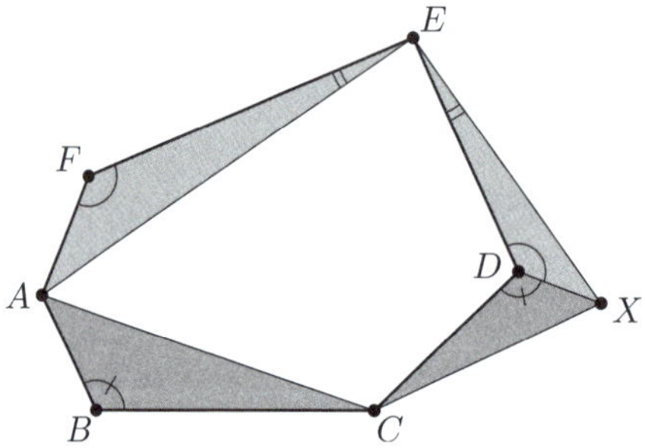

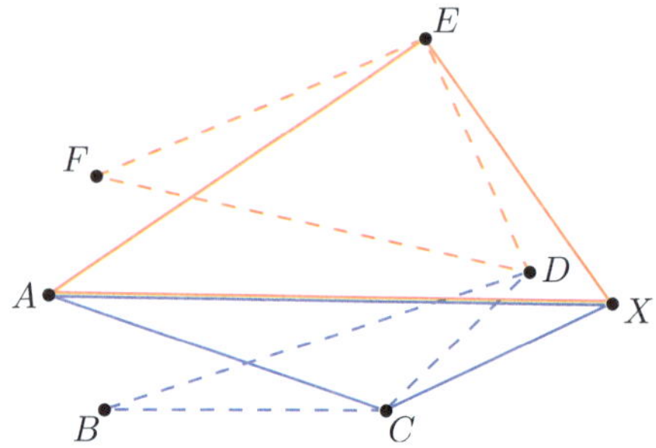

Since similarities come in pairs (see Proposition 1.45), we further obtain

$$\triangle EFD \sim \triangle EAX \quad \text{and} \quad \triangle CBD \sim \triangle CAX.$$

Finally, expressing the length AX from both the latter similarities yields

$$FD \cdot \frac{EA}{EF} = AX = BD \cdot \frac{CA}{CB}$$

which after regrouping terms proves the desired equality.

46. [Mathematical Reflections, Michal Rolínek] In acute scalene triangle ABC with orthocenter H, denote by α', β', and γ' the magnitudes of angles $180^\circ - \angle A$, $180^\circ - \angle B$, and $180^\circ - \angle C$, respectively. Points H_a, H_b, and H_c in the interior of triangle ABC satisfy

$$\begin{array}{rclrclrcl} \angle BH_aC &=& \alpha', & \angle CH_aA &=& \gamma', & \angle AH_aB &=& \beta', \\ \angle CH_bA &=& \beta', & \angle AH_bB &=& \alpha', & \angle BH_bC &=& \gamma', \\ \angle AH_cB &=& \gamma', & \angle BH_cC &=& \beta', & \angle CH_cA &=& \alpha'. \end{array}$$

Prove that the points H, H_a, H_b, H_c are concyclic.

First Proof. Let's first focus on point H_a and find out more about it. First of all, since $\angle BH_aC = 180^\circ - \angle A = \angle BHC$ (recall basic angles in a triangle from Proposition 1.35(c)), points B, C, H_a, and H lie on one circle and we may assume they lie on the circle in this order. Next, we note that we can angle-chase the magnitude of $\angle AH_aH$. Indeed,

$$\angle AH_aH = \angle AH_aB - \angle HH_aB = (180 - \angle B) - \underbrace{\angle HCB}_{=90^\circ - \angle B} = 90^\circ.$$

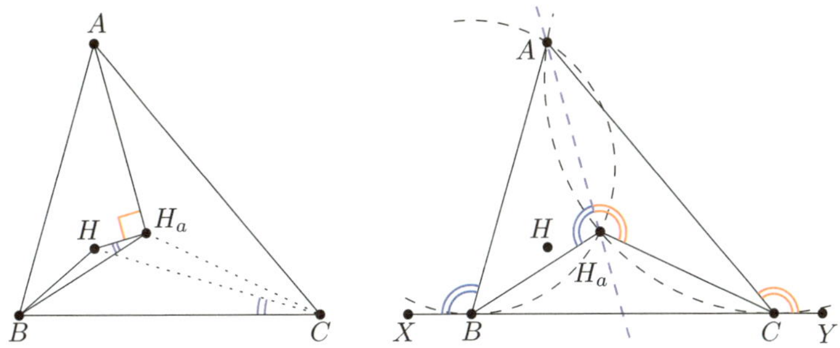

Although some could be satisfied with what we know about the point H_a, we will continue our investigation.

For notational purposes, let $X, Y \in BC$, such that points X, B, C, Y, lie on the line BC in this order. Since we have

$$\angle AH_aB = \angle ABX \quad \text{and} \quad \angle CH_aA = \angle YCA,$$

we infer that the line BC is tangent to the circumcircles of both triangle AH_aB and triangle AH_aC. But then the radical axis AH_a of the two circles intersects the common tangent BC at point M for which

$$MB^2 = MH_a \cdot MA = MC^2,$$

implying that AH_a is the median in triangle ABC.

Finally, since the median passes through the centroid G of triangle ABC, we may say that $\angle HH_aG = 90°$, implying that H_a lies on the circle with diameter HG.

Applying the same reasoning for H_b and H_c we obtain the result.

Second Proof. As in the first proof we find that B, C, H_a, and H lie on one circle and that $\angle HH_aA = 90°$. But now we invert about H (using standard notation for images $X \to X'$).

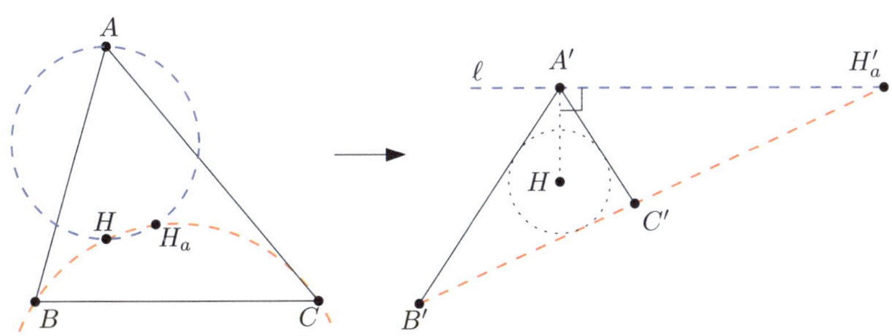

By Introductory Problem 44, H is the incenter of triangle $A'B'C'$. Also, the circle BHC goes to line $B'C'$ and thus $H'_a \in B'C'$. Finally, the circle with diameter AH goes to the line ℓ perpendicular to AH passing through A'. Thus $H'_a = B'C' \cap \ell$. But since ℓ is perpendicular to HA', the angle bisector in triangle $A'B'C'$, it is in fact the external angle bisector of $\angle B'A'C'$. Similarly, we find points H_b and H_c. The collinearity of H'_a, H'_b, and H'_c we are left to prove, already appeared in Introductory Problem 27(b).

Third Proof. (by Daniel Lasaosa) This time we will prove that the perpendicular bisectors of HH_a, HH_b, and HH_c are concurrent. As in the previous proofs we observe that H_a is the second intersection of the circle BHC and the circle with diameter AH (the other being H). Then, the perpendicular bisector of HH_a passes through the centers of both circles.

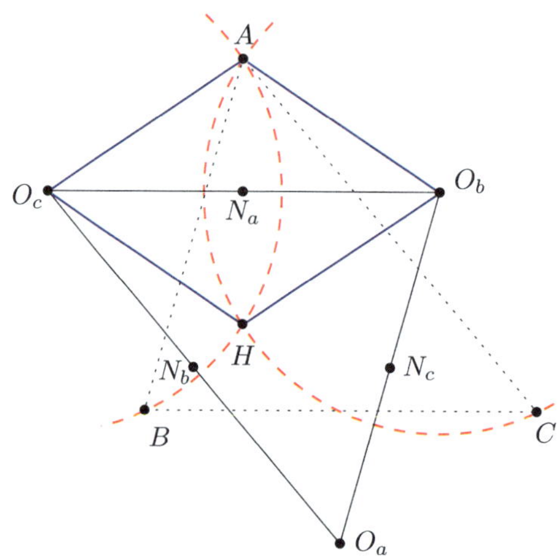

Therefore, we denote by O_a, O_b, and O_c the centers of circles BHC, CHA, and AHB and by N_a, N_b, and N_c the midpoints of segments HA, HB, HC. Now it suffices to prove that O_aN_a, O_bN_b, and O_cN_c are concurrent. But recalling that the circles CHA and AHB have equal radii (see Proposition 1.35(d)) we have $O_cA = O_cH = O_bH = O_bA$, implying that O_cAO_bH is a rhombus and therefore N_a is also the midpoint of O_cO_b. The desired point of concurrence is then the centroid of triangle $O_aO_bO_c$.

47. [IMO 2005 shortlist] Let ABC be an acute-angled triangle with $AB \neq AC$. Let H be the orthocenter of triangle ABC, and let M be the midpoint of the side BC. Let D be a point on the side AB and E a point on the side AC such that $AE = AD$ and the points D, H, E lie on

the same line. Prove that the line HM is perpendicular to the common chord of the circumscribed circles of the triangles ABC and ADE.

Proof. Denote by S the second intersection of the circumcircles of triangles ABC and ADE. Then S is the Miquel point of $BCED$ (see Theorem 1.49). Next, we exploit the condition $AD = AE$. Denote by B_0, C_0 the respective feet of altitudes in triangle ABC.

From $AD = AE$ we infer $\angle EDA = \angle AED = 90° - \frac{1}{2}\angle A$ and thus $\angle C_0HD = \angle EHB_0 = \frac{1}{2}\angle A$ which implies that DE is the angle bisector of BHC_0.

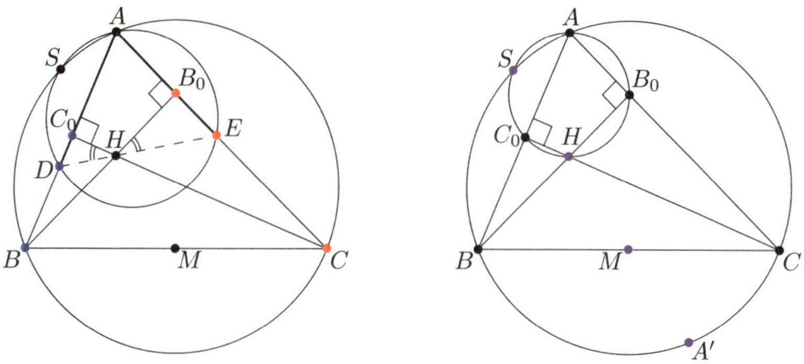

Since the triangles BHC_0 and CHB_0 are similar (quadrilateral BCB_0C_0 is cyclic) and points D and E correspond in this similarity, we have

$$\frac{BD}{DC_0} = \frac{CE}{EB_0}$$

and thus the spiral similarity centered at S that sends B to C and D to E maps also C_0 to B_0 implying that S lies also on the circumcircle of the triangle AC_0B_0. We continue in a figure without D and E.

Since AC_0HB_0 is cyclic, all the points A, S, C_0, H, and B_0 lie on a single circle with diameter AH. Denote by A' the point on the circumcircle of triangle ABC such that AA' is its diameter.

As $\angle ASH = 90° = \angle ASA'$, the points S, H, A' are collinear. At the same time, A' is the reflection of H about M (see Proposition 1.36) so the points H, M, A' are also collinear. Thus, the points S, H, M are collinear and $HM \perp AS$ as desired.

48. [Romania TST 1996] Let $ABCD$ be a cyclic quadrilateral. Draw all excenters of triangles ABC, BCD, CDA, and DAB. Show that these twelve points lie on the perimeter of a rectangle.

Proof. Recall that by Introductory Problem 30 the incenters of the triangles ABC, BCD, CDA, and DAB form a rectangle.

Denote by I_c, I_d the incenters of the triangles ABC, ABD, and by E_c, E_d their respective C- and D-excenters. We already know from the Big Picture (see Proposition 1.42(b)) that the midpoint M of the arc AB (not containing C) is the common center of the coinciding circles AI_cBE_c and AI_dBE_d. Since I_cE_c and I_dE_d are diameters of this circle, $E_cE_dI_cI_d$ is a rectangle.

Applying the very same reasoning to the arcs BC, CD, DA (not containing D, A, B, respectively) we learn that the four incenters together with eight of the excenters form some sort of a cross. It remains to prove that the last four excenters are the intersections of its outer sides.

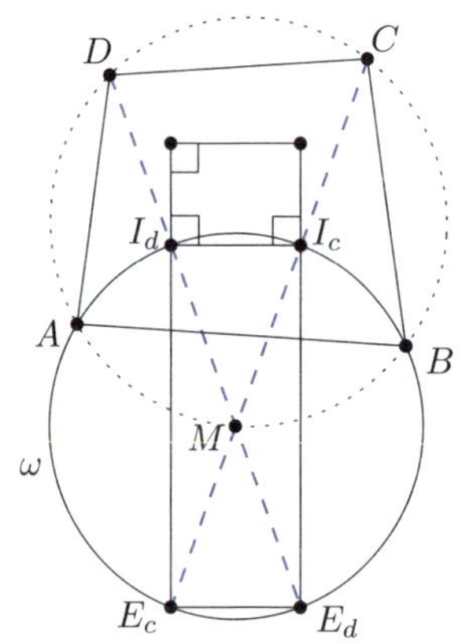

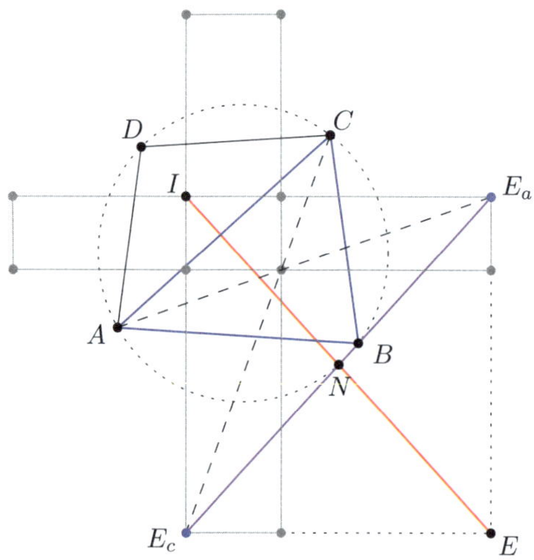

Let N be the midpoint of arc AC containing point B. Focusing on N with respect to triangle ACD we find it is the midpoint of IE, where I and E are the incenter and excenter of triangle ACD, respectively.

But at the same time, we note N is also the midpoint of E_aE_c, where E_a is the A-excenter of triangle ABC (again the Big Picture!).

Hence the diagonals E_aE_c and IE bisect each other at N and E_cEE_aI is a parallelogram. But since $\angle E_cIE_a = 90°$, it is in fact a rectangle. We are done.

49. [IMO 1998 shortlist] Let ABC be a triangle, H its orthocenter, O its circumcenter, and R its circumradius. Let D be the reflection of the point A across the line BC, let E be the reflection of the point B across

the line CA, and let F be the reflection of the point C across the line AB. Prove that the points D, E and F are collinear if and only if $OH = 2R$.

Proof. Recall that the center of the nine-point circle O_9 of the triangle ABC is the midpoint of OH (see Theorem 1.37). Hence $OH = 2R$ holds if and only if O_9 belongs to the circumcircle of triangle ABC. This rewording seems more promising.

A point on a circle and a collinearity should remind us of the Simson line (see Proposition 1.44).

If we denote by X, Y, Z the projections of O_9 onto the lines BC, CA, AB, then O_9 belongs to the circumcircle of triangle ABC if and only if the points X, Y, Z lie on a single line.

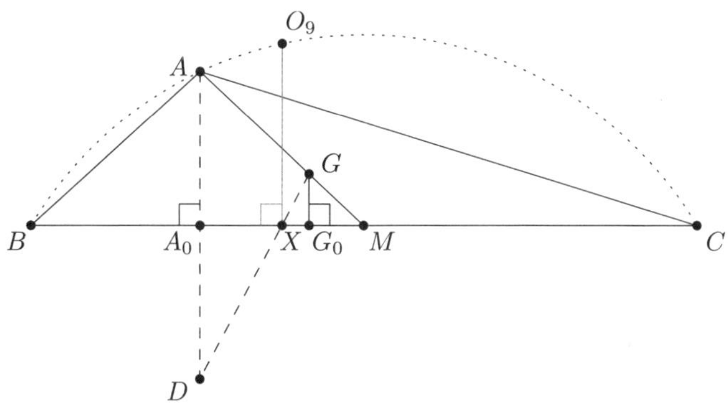

Now we will prove that the points D, E, F are the images of X, Y, Z, respectively, under some very particular homothety. Then the result will follow since under homothety, the images of three points lie on a single line if and only if the initial points do so.

The center of this mysterious homothety will be the centroid G of triangle ABC and the factor will be 4. Once we manage to guess it, the rest can be done by many approaches.

For instance, let M be the midpoint of BC and A_0 the foot of A-altitude.

Since both A_0 and M lie on the nine-point circle of triangle ABC, we have $O_9A_0 = O_9M$ and so point X is the midpoint of A_0M. Menelaus' Theorem applied for triangle AA_0M and points D, X, and G yields

$$\frac{AD}{DA_0} \cdot \frac{A_0X}{XM} \cdot \frac{MG}{GA} = \frac{2}{1} \cdot \frac{1}{1} \cdot \frac{1}{2} = 1.$$

Thus, the points G, X, D are collinear. Finally, let G_0 be the projection of G to BC. As G "trisects" the median and $AA_0 = A_0D$, we obtain $GX/XD = GG_0/AA_0 = \frac{1}{3}$. We are done.

50. [IMO 2006 shortlist] Points A_1, B_1, C_1 are chosen on the sides BC, CA, AB of a triangle ABC, respectively. The circumcircles of triangles AB_1C_1, BC_1A_1, CA_1B_1 intersect the circumcircle ω of triangle ABC for the second time at points A_2, B_2, C_2, respectively. Points A_3, B_3, C_3 are symmetric to A_1, B_1, C_1 with respect to the midpoints of the sides BC, CA, AB, respectively. Prove that the triangles $A_2B_2C_2$ and $A_3B_3C_3$ are similar.

Proof. First of all, we identify A_2 as the center of spiral similarity $\mathcal{S}(A_2, k, \angle A)$ which (for some k) takes C_1 to B_1 and B to C (see Proposition 1.47). Then it takes BC_1 to CB_1, thus its factor k equals

$$k = \frac{CB_1}{BC_1}.$$

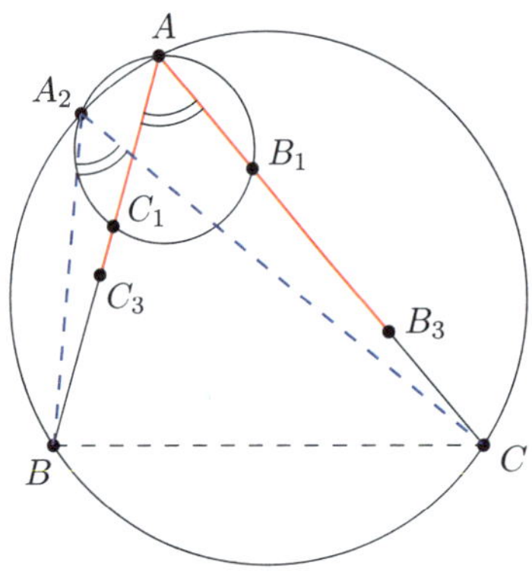

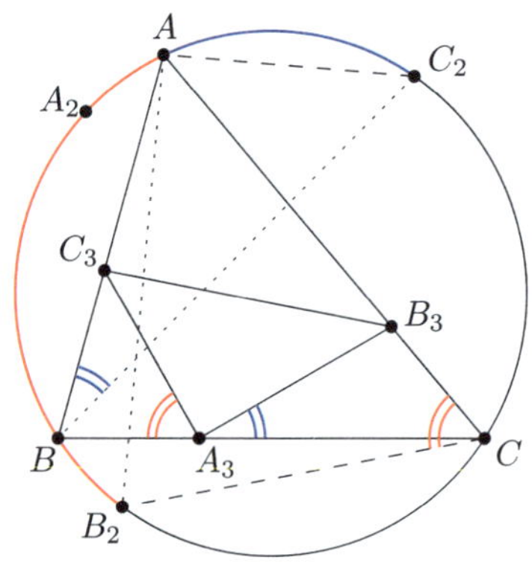

This gives us a chance to use the definition of B_3 and C_3, as we have $BC_1 = AC_3$ and $CB_1 = AB_3$ and thus also $k = AB_3/AC_3$. Now the vital observation is that triangle AB_3C_3 has the very shape that is produced by spiral similarity $\mathcal{S}$! Therefore, we have $\triangle AB_3C_3 \sim \triangle A_2CB$ (SAS). Similar argument shows $\triangle BC_3A_3 \sim \triangle B_2AC$ and $\triangle CA_3B_3 \sim \triangle C_2BA$.

The rest is easy, since we can forget points A_1, B_1, C_1 and represent angles in triangle AB_3C_3 (and the other two) as some arcs of ω. Indeed, writing this down in the language of directed angles gives

$$\begin{aligned}
\angle(C_3A_3, A_3B_3) &= \angle(C_3A_3, BC) + \angle(BC, A_3B_3) \\
&= \angle(AC, CB_2) + \angle(C_2B, BA) \\
&= \angle(AA_2, A_2B_2) + \angle(C_2A_2, A_2A) = \angle(C_2A_2, A_2B_2)
\end{aligned}$$

and the conclusion follows from analogous arguments.

51. [IMO 2002 shortlist] The incircle ω of the acute-angled triangle ABC is tangent to its side BC at a point K. Let AD be an altitude of triangle ABC, and let M be its midpoint. If N is the common point of the circle ω and the line KM (distinct from K), then prove that the incircle ω and the circumcircle ω' of triangle BCN are tangent to each other at the point N.

First Proof. If $b = c$, the problem is trivial. Hence we may assume $b > c$. We introduce point $N' \in \omega$ such that circle through BCN' is tangent to ω. Being clueless as to what we should do with the midpoint of AD, we choose a computational approach to show that $N = N'$. Observing that both distances DM and DK are approachable in terms of x, y, z, we decide to prove that

$$\tan \angle NKB = \tan \angle N'KB.$$

We plan to express both sides in x, y, and z and then easily compare. As mentioned, for the left-hand side it is easy:

$$\tan \angle NKB = \frac{MD}{DK} = \frac{AD/2}{BK - BD} = \frac{K}{a(y - c\cos \angle B)}$$
$$= \frac{2K}{2y(y+z) - 2ac\cos \angle B},$$

where K denotes the area of triangle ABC. Even though we used more triangle elements than just x, y, and z this form suffices for now.

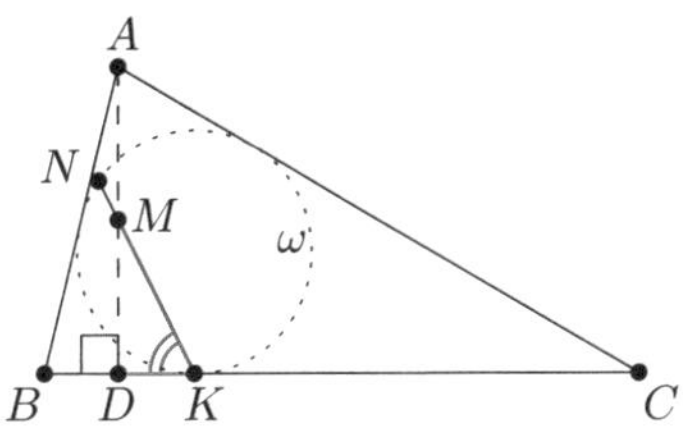

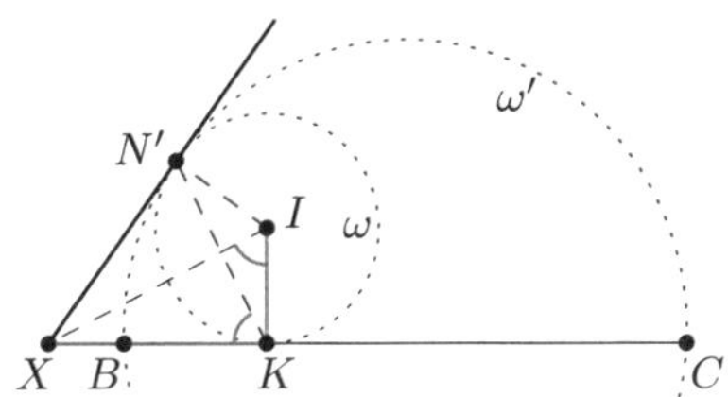

For the right-hand side we need more thought. The good thing is, we can erase point A. Draw the common tangent of ω and ω' at N' and denote its intersection with BC by X. Also, let I be the center of ω. Since $XKIN'$ is a cyclic kite, we have $\angle N'KB = \angle XIK$, hence

$$\tan \angle N'KB = \tan \angle XIK = \frac{XK}{KI}.$$

Also, by Power of a Point we have

$$XB \cdot XC = XN'^2 = XK^2,$$

from which we can find

$$(XK - y)(XK + z) = XK^2 \quad \text{and} \quad XK = \frac{yz}{z - y}.$$

At this point, we are only left to do some routine algebra, since using xyz formulas (see Proposition 1.8) we can express everything in x, y, and z. We will just ease our lives a bit by clever use of area formulas:

$$\tan \angle N'KB = \frac{yz}{r(z-y)} = \frac{syz}{K(z-y)} = \frac{2K}{2x(z-y)}.$$

Now it suffices to compare the denominators. After using the Law of Cosines, we are left to prove

$$2x(z-y) = 2y^2 + 2yz + (x+z)^2 - (y+x)^2 - (y+z)^2,$$

which is immediate after expanding, since as we can see, the right-hand side simplifies significantly.

Thus, we have proved $N = N'$ and the problem is solved.

Second Proof. A synthetic approach is not only possible, but also very beautiful. Again we work with the incenter I of triangle ABC and we draw the tangent to ω at N and denote its intersection with BC by X. By Power of a Point, we need to prove that $XN^2 = XB \cdot XC$. Note that the point $P = KN \cap IX$ is the midpoint of KN and also that $XI \perp KN$. From right triangle INX, we learn (see Introductory Problem 2) $XN^2 = XP \cdot XI$. Thus, we need to prove that point P lies on the circumcircle of triangle BIC.

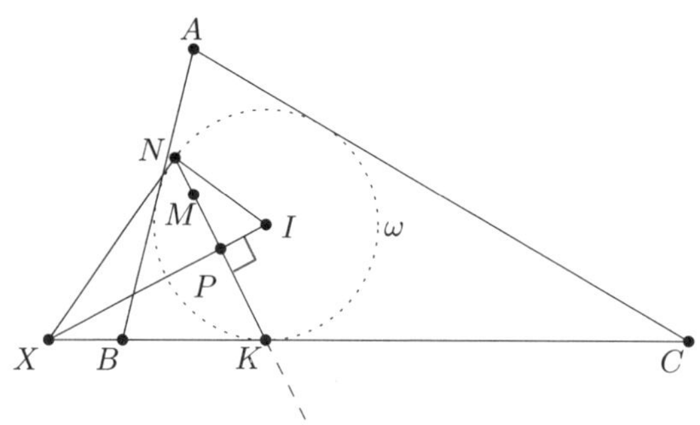

Looking at the right angle KPI we decide to introduce the A-excenter E of triangle ABC, since it is antipodal to I in circle BIC as we know from the Big Picture (see Proposition 1.42). But now we only need to prove that E lies on the line KM! We have reduced the problem to a simpler one, but there is still work ahead of us.

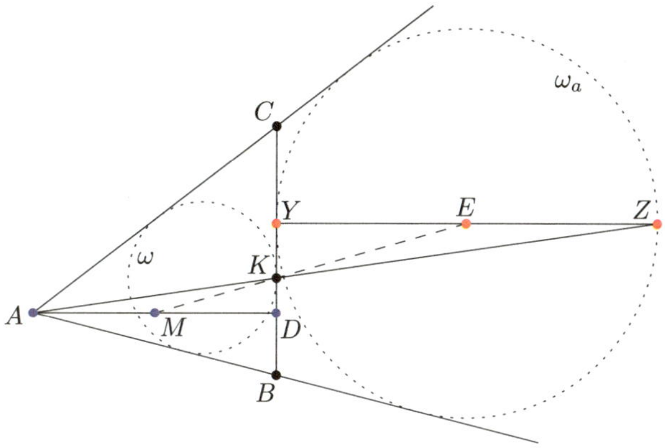

Let Y be the point of contact of the A-excircle ω_a with BC and let YZ be a diameter of ω_a. We place BC vertically and consider homothety with center A which sends ω to ω_a. Since K and Z are the "rightmost" points on the respective circles, they correspond in the homothety and thus are collinear with A. Finally, this means that K is the center of a homothety which takes triangle ADK to triangle ZYK and thus the midpoint of AD is taken to the midpoint of ZY, i.e. M is taken to E. This proves the collinearity of M, K, and E and we may conclude.

52. [Sawayama's Lemma] Let ABC be a triangle inscribed in the circle ω. Point D is chosen on the side BC. Circle ω_1 is tangent to the segment BD at K, to the segment AD at L and to ω at T. Prove that the line KL passes through the incenter I of the triangle ABC.

Proof. (Inspired by ideas of Jean-Louis Ayme[7]) Without loss of generality assume $\angle DAC < \frac{1}{2}\angle A$ and place BC horizontally.

To make use of the tangency of the circles ω and ω_1, denote by K' the second intersection of TK with ω. The homothety with center T which takes ω_1 to ω, then takes K to K', thus K' is the "bottom" point of ω i.e. the midpoint of arc BC (not containing A). A connection with the incenter emerges. Draw the bisector AK' of $\angle A$.

Instead of dealing with I, let J be the intersection of AK' and KL. We will prove that J in fact coincides with I.

Since J lies on the angle bisector of $\angle A$, it suffices to prove it has the expected distance from K', i.e. that $K'J^2 = K'B^2$ (recall Proposition 1.38).

[7] Jean-Louis Ayme is a French professor of mathematics and a passionate geometer. It is not an overstatement to call him a world leading expert in synthetic geometry.

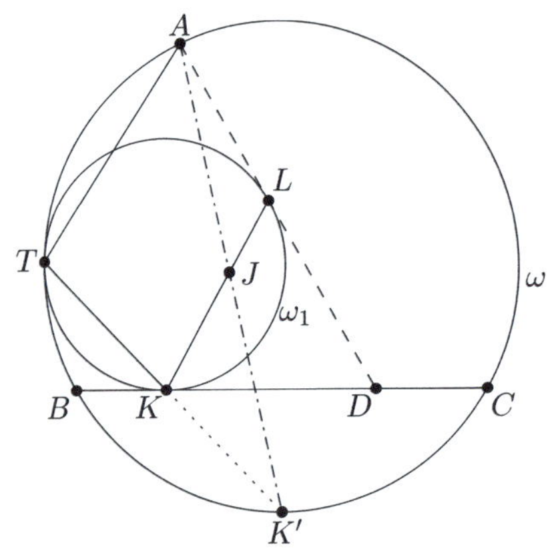

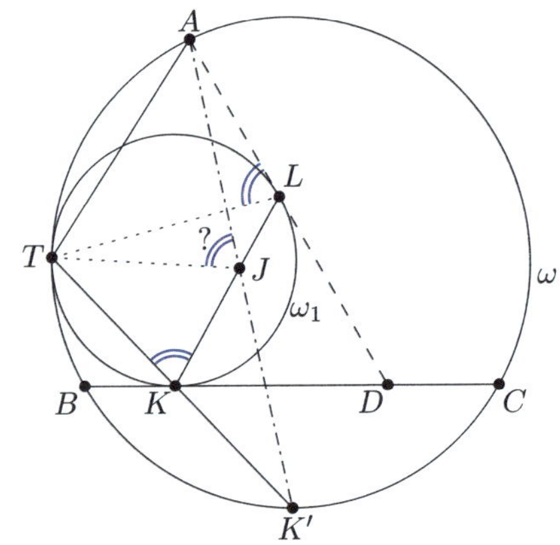

As the latter equals $K'K \cdot K'T$ (see Shooting Lemma – Proposition 1.40(b)), by Power of a Point it is enough to prove that the circumcircle of TKJ is tangent to $K'A$, or in other words that $\angle JKT = \angle AJT$.

Since $\angle JKT$ subtends arc LT on ω_1, it is equal to $\angle ALT$. The whole problem therefore reduces to proving that quadrilateral $ATJL$ is cyclic, which is straightforward, since we may erase points B, D and C. We offer two approaches.

First approach. Let ℓ be the common tangent of ω_1 and ω. Since the angle $\angle TLK$ inscribed in ω_1 and angle $\angle TAK'$ inscribed in ω are both equal to the same angle by ℓ, quadrilateral $ATJL$ is cyclic.

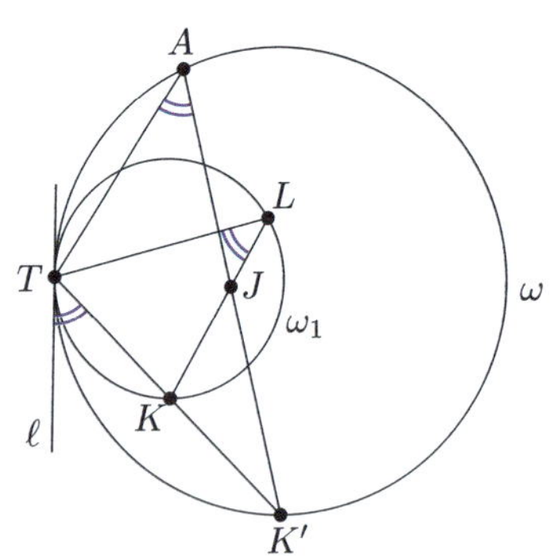

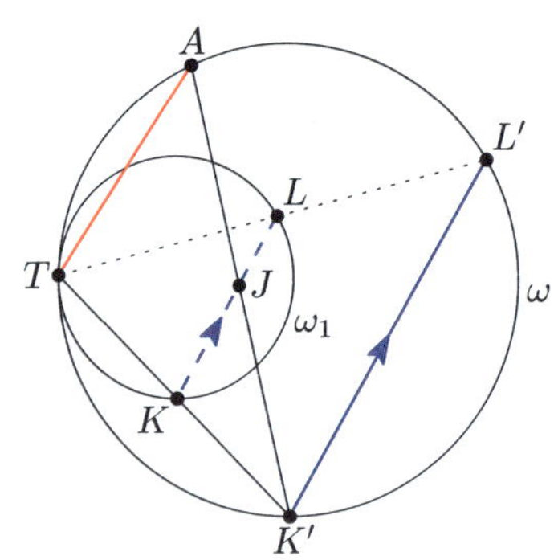

Second approach. Let L' be the second intersection of TL and ω. The homothety centered at T which sends ω_1 to ω maps KL to $K'L'$, so $KL \parallel K'L'$. Looking at angle between lines AJ and TL, line $K'L'$ is antiparallel to AT and thus so is JL.

Remark. This problem together with Introductory Problem 38 establishes the celebrated Sawayama[8]-Thébault[9]'s Theorem (Can you recognize the configuration?) which states that in the following diagram, lines KL, MN, and I_1I_2 are concurrent at the incenter I of triangle ABC.

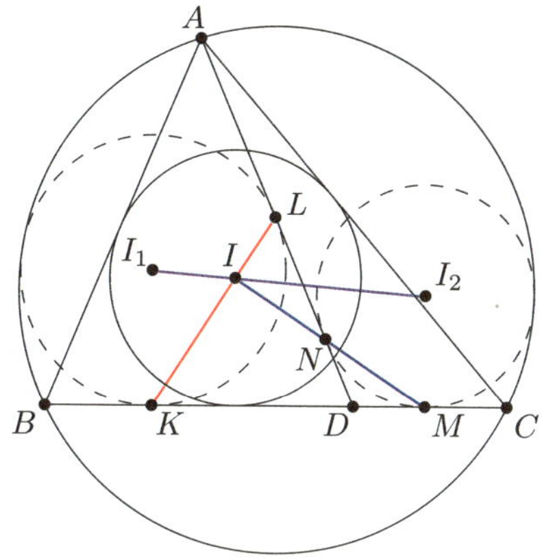

53. [IMO 2008] Let $ABCD$ be a convex quadrilateral with BA different from BC. Denote the incircles of triangles ABC and ADC by ω_1 and ω_2, respectively. Suppose that there exists a circle ω tangent to ray BA beyond A and to the ray BC beyond C, which is also tangent to the lines AD and CD. Prove that the common external tangents to ω_1 and ω_2 intersect on ω.

Proof. Place AC horizontally with B above it.

First, recall that since $ABCD$ has "escribed" circle ω, the incircles ω_1, ω_2 of the triangles ABC, ADC are tangent to the diagonal AC at two symmetric points (see Introductory Problem 51(c)).

Thus, if we denote them by P, Q, respectively, then Q is the point of contact of the B-excircle of triangle ABC and similarly, P is the point of contact of the D-excircle of triangle ADC, both with AC (recall Proposition 1.7).

Hence the line BQ passes through the "top" point of ω_1 (denote it by Q'), and DP passes through the "bottom" point of ω_2, which we denote by P' (see Proposition 1.30).

The intersection of the common external tangents to ω_1 and ω_2 is nothing but the center of the positive homothety $\mathcal{H}$ that maps ω_1 to ω_2 (see Proposition 1.29). Forget the tangents.

[8]Yusaburo Sawayama (1860–1936) was a military instructor in Tokyo with genuine interest in geometry.

[9]Victor Thébault (1882–1960) was a renowned French geometer.

Let X be the "top" point of ω. We will prove that X is the center of $\mathcal{H}$.

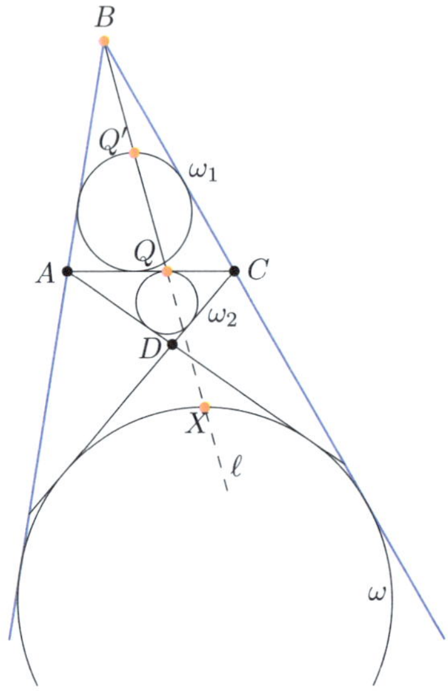

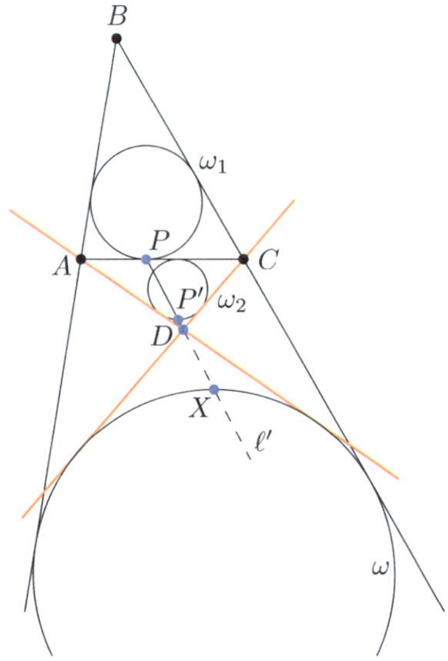

First, we focus on the line passing through B, Q', and Q. Denote it by ℓ.

As Q' and Q are the corresponding points on ω_1, ω_2 (namely their "top" points), line ℓ passes through the center of $\mathcal{H}$. However, since both ω_1 and ω are inscribed in angle ABC, and ℓ intersects ω_1 at its "top" point Q', it intersects ω at its "top" point (i.e. X) too.

Similarly, denote by ℓ' the line passing through D, P', and P.

Then P and P' also correspond under $\mathcal{H}$ (they are the "bottom" points on ω_1, ω_2), hence ℓ' passes through the center of $\mathcal{H}$. At the same time, the homothety centered at D which sends ω_2 to ω maps the "bottom" point P' of ω_2 to the "top" point X of ω. Thus, ℓ' also passes through X.

Finally, from $AB \neq BC$ we infer that ℓ and ℓ' do not coincide. As the center of $\mathcal{H}$ lies on both of them, it has to be X.

54. [Kolmogorov cup 2014, Michal Rolínek] Let XVY be a fixed angle with magnitude φ and P a point inside of it. A variable angle APB with $A \in VX$ and $B \in VY$ and fixed magnitude $\psi > \varphi$ and $2\psi - \varphi \neq 180°$ revolves around P. Show that there exists point Q in the plane ($Q \neq P$, $Q \neq V$) such that the angle AQB is fixed.

Proof. We will handle the various configurations all at once with directed angles mod 180°. We can set $\angle(XV, VY) = \varphi$ and $\angle(AP, PB) = \psi$. Let A_∞ be a point on VX that that $\angle(PA_\infty, VY) = \psi$ and similarly let B_∞ be point on VY such that $\angle(VX, PB_\infty) = \psi$ (i.e. points corresponding to limit positions of the angle APB).

Let Q be a point such that $A_\infty PB_\infty Q$ is a (possibly degenerate) parallelogram. We will prove that Q has the desired property.

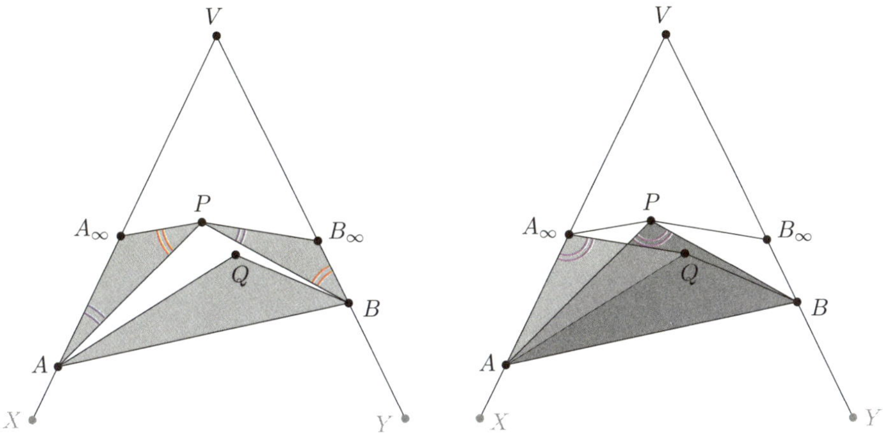

For some fixed A and B, we have

$$\begin{aligned}\angle(A_\infty P, AP) &= \angle(A_\infty P, VB) + \angle(VB, AP) = \\ &= \angle(AP, PB) + \angle(VB, AP) = \angle(VB, PB)\end{aligned}$$

and similarly we derive that $\angle(BP, PB_\infty) = \angle(PA, AV)$. Hence $\triangle AA_\infty P \sim \triangle PB_\infty B$. Then

$$\frac{AA_\infty}{A_\infty Q} = \frac{AA_\infty}{PB_\infty} = \frac{PA}{PB}$$

and since also $\angle(AA_\infty, A_\infty Q) = \angle(AA_\infty, PB_\infty) = \psi = \angle(PA, PB)$, we have $\triangle APB \sim \triangle AA_\infty Q$ (directly). Considering spiral similarity centered at A which sends $\triangle APB$ to $\triangle AA_\infty Q$ makes us notice another pair of similar triangles $\triangle AQB \sim \triangle AA_\infty P$. But now we are done as

$$\angle(AQ, QB) = \angle(AA_\infty, A_\infty P) = \varphi + \angle(BV, A_\infty P) = \varphi - \psi,$$

which is fixed. It remains to note that $\varphi - \psi \neq \varphi$ and $\varphi - \psi \neq \psi$ so indeed $Q \neq V$ and $Q \neq P$ as desired.

Further Reading

1. Altschiller-Court, N., *College Geometry, an Introduction to the Modern Geometry of the Triangle and the Circle*, Dover publications, 2007.
2. Andreescu, T.; Feng, Z., *101 Problems in Algebra from the Training of the USA IMO Team*, Australian Mathematics Trust, 2001.
3. Andreescu, T.; Feng, Z., *102 Combinatorial Problems from the Training of the USA IMO Team*, Birkhäuser, 2002.
4. Andreescu, T.; Feng, Z., *103 Trigonometry Problems from the Training of the USA IMO Team*, Birkhäuser, 2004.
5. Andreescu, T.; Andrica, D.; Feng, Z., *104 Number Theory Problems from the Training of the USA IMO Team*, Birkhäuser, 2006.
6. Andreescu, T.; Feng, Z.; Loh, P., *USA and International Mathematical Olympiads 2004*, Mathematical Association of America, 2005.
7. Andreescu, T.; Feng, Z., *USA and International Mathematical Olympiads 2003*, Mathematical Association of America, 2004.
8. Andreescu, T.; Feng, Z., *USA and International Mathematical Olympiads 2002*, Mathematical Association of America, 2003.
9. Andreescu, T.; Feng, Z., *USA and International Mathematical Olympiads 2001*, Mathematical Association of America, 2002.
10. Andreescu, T.; Feng, Z., *USA and International Mathematical Olympiads 2000*, Mathematical Association of America, 2001.
11. Andreescu, T.; Feng, Z.; Lee, G.; Loh, P., *Mathematical Olympiads: Problems and Solutions from around the World, 2001–2002*, Mathematical Association of America, 2004.
12. Andreescu, T.; Feng, Z.; Lee, G., *Mathematical Olympiads: Problems and Solutions from around the World, 2000–2001*, Mathematical Association of America, 2003.

13. Andreescu, T.; Feng, Z., *Mathematical Olympiads: Problems and Solutions from around the World, 1999–2000*, Mathematical Association of America, 2002.

14. Andreescu, T.; Feng, Z., *Mathematical Olympiads: Problems and Solutions from around the World, 1998–1999*, Mathematical Association of America, 2000.

15. Andreescu, T.; Kedlaya, K., *Mathematical Contests 1997–1998: Olympiad Problems from around the World, with Solutions*, American Mathematics Competitions, 1999.

16. Andreescu, T.; Kedlaya, K., *Mathematical Contests 1996–1997: Olympiad Problems from around the World, with Solutions*, American Mathematics Competitions, 1998.

17. Andreescu, T.; Kedlaya, K.; Zeitz, P., *Mathematical Contests 1995–1996: Olympiad Problems from around the World, with Solutions*, American Mathematics Competitions, 1997.

18. Andreescu, T.; Enescu, B., *Mathematical Olympiad Treasures*, 2nd edition, Birkhäuser, 2011.

19. Andreescu, T.; Gelca, R., *Mathematical Olympiad Challenges*, 2nd edition, Birkhäuser, 2009.

20. Andreescu, T.; Andrica, D.; Cucurezeanu, I., *An Introduction to Diophantine Equations: A Problem-Based Approach*, Birkhäuser, 2010.

21. Andreescu, T.; Andrica, D., *360 Problems for Mathematical Contests*, GIL Publishing House, 2003.

22. Andreescu, T.; Andrica, D., *Complex Numbers from A to Z*, Birkhäuser, 2004.

23. Andreescu, T.; Feng, Z., *A Path to Combinatorics for Undergraduate Students: Counting Strategies*, Birkhäuser, 2003.

24. Andreescu, T.; Andrica, D., *Number Theory - A Problem Solving Approach*, Birkhäuser, 2009.

25. Ayme, J-L., *Sawayama and Thebault's Theorem*, Forum Geom. 3, 225-229, 2003.

26. Coxeter, H. S. M.; Greitzer, S. L., *Geometry Revisited*, New Mathematical Library, Vol. 19, Mathematical Association of America, 1967.

27. Coxeter, H. S. M., *Non-Euclidean Geometry*, The Mathematical Association of America, 1998.

28. Doob, M., *The Canadian Mathematical Olympiad 1969–1993*, University of Toronto Press, 1993.

29. Engel, A., *Problem-Solving Strategies*, Problem Books in Mathematics, Springer, 1998.

30. Fomin, D.; Kirichenko, A., *Leningrad Mathematical Olympiads 1987–1991*, MathPro Press, 1994.

31. Fomin, D.; Genkin, S.; Itenberg, I., *Mathematical Circles*, American Mathematical Society, 1996.

32. Gelca, R.; Andreescu, T., *Putnam and Beyond*, Springer, 2007.

33. Graham, R. L.; Knuth, D. E.; Patashnik, O., *Concrete Mathematics*, Addison-Wesley, 1989.

34. Greitzer, S. L., *International Mathematical Olympiads, 1959–1977*, New Mathematical Library, Vol. 27, Mathematical Association of America, 1978.

35. Holton, D., *Let's Solve Some Math Problems*, A Canadian Mathematics Competition Publication, 1993.

36. Kazarinoff, N. D., *Geometric Inequalities*, New Mathematical Library, Vol. 4, Random House, 1961.

37. Kedlaya, K; Poonen, B.; Vakil, R., *The William Lowell Putnam Mathematical Competition 1985–2000*, The Mathematical Association of America, 2002.

38. Klamkin, M., *International Mathematical Olympiads, 1978–1985*, New Mathematical Library, Vol. 31, Mathematical Association of America, 1986.

39. Klamkin, M., *USA Mathematical Olympiads, 1972–1986*, New Mathematical Library, Vol. 33, Mathematical Association of America, 1988.

40. Kürschák, J., *Hungarian Problem Book, volumes I & II*, New Mathematical Library, Vols. 11 & 12, Mathematical Association of America, 1967.

41. Kuczma, M., *144 Problems of the Austrian–Polish Mathematics Competition 1978–1993*, The Academic Distribution Center, 1994.

42. Kuczma, M., *International Mathematical Olympiads 1986–1999*, Mathematical Association of America, 2003.

43. Larson, L. C., *Problem-Solving Through Problems*, Springer-Verlag, 1983.

44. Lausch, H. *The Asian Pacific Mathematics Olympiad 1989–1993*, Australian Mathematics Trust, 1994.

45. Liu, A., *Chinese Mathematics Competitions and Olympiads 1981–1993*, Australian Mathematics Trust, 1998.

46. Liu, A., *Hungarian Problem Book III*, New Mathematical Library, Vol. 42, Mathematical Association of America, 2001.

47. Lozansky, E.; Rousseau, C. *Winning Solutions*, Springer, 1996.

48. Mitrinovic, D. S.; Pecaric, J. E.; Volonec, V. *Recent Advances in Geometric Inequalities*, Kluwer Academic Publisher, 1989.

49. Mordell, L.J., *Diophantine Equations*, Academic Press, London and New York, 1969.

50. Niven, I., Zuckerman, H.S., Montgomery, H.L., *An Introduction to the Theory of Numbers*, Fifth Edition, John Wiley & Sons, Inc., New York, Chichester, Brisbane, Toronto, Singapore, 1991.

51. Prasolov, V.V., *Problems in Plane Geometry*, Fifth Edition, Moscow textbooks, 2006.

52. Savchev, S.; Andreescu, T. *Mathematical Miniatures*, Anneli Lax New Mathematical Library, Vol. 43, Mathematical Association of America, 2002.

53. Sharygin, I. F., *Problems in Plane Geometry*, Mir, Moscow, 1988.

54. Sharygin, I. F., *Problems in Solid Geometry*, Mir, Moscow, 1986.

55. Shklarsky, D. O; Chentzov, N. N; Yaglom, I. M., *The USSR Olympiad Problem Book*, Freeman, 1962.

56. Slinko, A., *USSR Mathematical Olympiads 1989–1992*, Australian Mathematics Trust, 1997.

57. Szekely, G. J., *Contests in Higher Mathematics*, Springer-Verlag, 1996.

58. Tattersall, J.J., *Elementary Number Theory in Nine Chapters*, Cambridge University Press, 1999.

59. Taylor, P. J., *Tournament of Towns 1980–1984*, Australian Mathematics Trust, 1993.

60. Taylor, P. J., *Tournament of Towns 1984–1989*, Australian Mathematics Trust, 1992.

61. Taylor, P. J., *Tournament of Towns 1989–1993*, Australian Mathematics Trust, 1994.

62. Taylor, P. J.; Storozhev, A., *Tournament of Towns 1993–1997*, Australian Mathematics Trust, 1998.

63. Yaglom, I. M., *Geometric Transformations*, New Mathematical Library, Vol. 8, Random House, 1962.

64. Yaglom, I. M., *Geometric Transformations II*, New Mathematical Library, Vol. 21, Random House, 1968.

65. Yaglom, I. M., *Geometric Transformations III*, New Mathematical Library, Vol. 24, Random House, 1973.

Other Books from XYZ Press

1. Andreescu, T., *105 Algebra Problems from the AwesomeMath Summer Program*, 2013.
2. Andreescu, T.; Rolínek, M.; Tkadlec, J, *106 Geometry Problems from the AwesomeMath Summer Program*, 2013.
3. Andreescu, T.; Kane, J., *Purple Comet Math Meet! - the first ten years*, 2013.
4. Andreescu, T.; Dospinescu, G., *Straight from the Book*, 2012.
5. Andreescu, T.; Boreico, I.; Mushkarov, O.; Nikolov, N., *Topics in Functional Equations*, 2012.
6. Andreescu, T., *Mathematical Reflections - the next two years*, 2012.
7. Andreescu, T., *Mathematical Reflections - the first two years*, 2011.
8. Andreescu, T.; Dospinescu, G., *Problems from the Book*, 2008.

Index

Printed by "Combinatul Poligrafic"
Com. nr. 80975